CAMBRIDGE LIBRARY COLLECTION

Books of enduring scholarly value

Physical Sciences

From ancient times, humans have tried to understand the workings of the world around them. The roots of modern physical science go back to the very earliest mechanical devices such as levers and rollers, the mixing of paints and dyes, and the importance of the heavenly bodies in early religious observance and navigation. The physical sciences as we know them today began to emerge as independent academic subjects during the early modern period, in the work of Newton and other 'natural philosophers', and numerous sub-disciplines developed during the centuries that followed. This part of the Cambridge Library Collection is devoted to landmark publications in this area which will be of interest to historians of science concerned with individual scientists, particular discoveries, and advances in scientific method, or with the establishment and development of scientific institutions around the world.

Memoir of the Life and Labors of the Rev. Jeremiah Horrox

Jeremiah Horrox (1618–41) was one of the most interesting astronomers Britain has ever produced, and his tragically early death deprived the field of one of its most brilliant talents. In his short life he achieved much, having mastered the current state of astronomy at Cambridge University and going on to make important new calculations about the diameter and position of known planets, moons and stars. In the 1660s and 70s several prominent scientists, including Huygens, Newton and Flamsteed, took an interest in Horrox's discoveries and published his surviving treatises. This memoir of 1859 was part of a Victorian revival of interest in Horrox. It includes a translation of his major work, *Venus in Sole Visa*, a draft of a treatise on the transit of Venus, in which he describes the conjunction of Venus with the sun, which he correctly calculated and observed in 1639.

Cambridge University Press has long been a pioneer in the reissuing of out-of-print titles from its own backlist, producing digital reprints of books that are still sought after by scholars and students but could not be reprinted economically using traditional technology. The Cambridge Library Collection extends this activity to a wider range of books which are still of importance to researchers and professionals, either for the source material they contain, or as landmarks in the history of their academic discipline.

Drawing from the world-renowned collections in the Cambridge University Library, and guided by the advice of experts in each subject area, Cambridge University Press is using state-of-the-art scanning machines in its own Printing House to capture the content of each book selected for inclusion. The files are processed to give a consistently clear, crisp image, and the books finished to the high quality standard for which the Press is recognised around the world. The latest print-on-demand technology ensures that the books will remain available indefinitely, and that orders for single or multiple copies can quickly be supplied.

The Cambridge Library Collection will bring back to life books of enduring scholarly value (including out-of-copyright works originally issued by other publishers) across a wide range of disciplines in the humanities and social sciences and in science and technology.

Memoir of the Life and Labors of the Rev. Jeremiah Horrox

To Which is Appended a Translation of his Celebrated Discourse Upon the Transit of Venus Across the Sun

EDITED BY ARUNDELL BLOUNT WHATTON

CAMBRIDGE UNIVERSITY PRESS

Cambridge, New York, Melbourne, Madrid, Cape Town, Singapore,
São Paolo, Delhi, Dubai, Tokyo

Published in the United States of America by Cambridge University Press, New York

www.cambridge.org
Information on this title: www.cambridge.org/9781108014410

This edition first published 1859
This digitally printed version 2010

ISBN 978-1-108-01441-0 Paperback

MEMOIR

OF THE

LIFE AND LABORS

OF THE

REV. JEREMIAH HORROX,

Curate of Hoole, near Preston;

TO WHICH IS APPENDED

A TRANSLATION OF HIS CELEBRATED DISCOURSE

UPON THE

Transit of Venus across the Sun.

BY THE

REV. ARUNDELL BLOUNT WHATTON, B.A., LL.B.

LONDON:

WERTHEIM, MACINTOSH, AND HUNT,

24, PATERNOSTER ROW,

AND 23, HOLLES STREET, CAVENDISH SQUARE.

MDCCCLIX.

In Memoriam

PATRIS DILECTISSIMI

GUL: ROB: WHATTON, F.R.S.: F.S.A., ETC.,

VIRI LITERIS HUMANIORIBUS EXIMIE ERUDITI,

HAS EGREGII ADOLESCENTIS RELIQUIAS,

QUEM VIVUS IPSE EST MIRATUS,

CUM QUO MORTUUS, FAS EST CREDERE, CONSORS,

COLLIGENDAS ET IN LUMEN PROFERENDAS CURAVIT

FILIUS AMANTISSIMUS

A. B. W.

CONTENTS.

PREFACE.

WHEN my father was engaged in writing the Biographical department of the history of Lancashire, he was naturally led to consider the merits of JEREMIAH HORROX, the youthful astronomer of that county; and he was so much impressed with his distinguished scientific attainments that, finding it impossible from want of space to do him justice in those pages, he proposed on some future occasion to publish his life in a separate form. Accordingly, he ascertained the precise value of his discoveries, and gathered together much interesting detail connected with his personal history; and he also set about preparing a translation of his celebrated Treatise upon the transit of Venus over the Sun. But he did not live to complete this work. It would appear that much material had been accumulated, but that the arrangement of it had not even been commenced. To him however belongs the credit of being the first and only person who has undertaken to supply what is acknowledged to be a deficiency in the literature of our country; and there can be no doubt that, if his life had been spared a little longer, he would have produced a most interesting and instructive volume. Professor Rigaud, of Oxford, who was his friend and associate in these pursuits, says in his "Correspondence of Scientific Men of the 17th century," that "the late W. R. Whatton, Esquire, had made considerable collections for a life of Horrox, which he intended to have prefixed to a new edition of the *Venus in sole visa*, when death in 1835 deprived the world of the fruit of his inquiries."

Since then no further attempt of this kind has been made to recognize the merits, or to perpetuate the memory of Horrox. Of late years, however, his name, associated with the names of other persons of distinction, has been brought before the public from time to time by various speakers at literary and scientific meetings, especially in Lancashire. Thus, in an address delivered in Liverpool on the celebration of the centenary of the birthday of Roscoe, the Rev. Dr. Hume says: "neither is Roscoe the first man of high intellectual attainments that Liverpool has numbered among her sons. More than two centuries have elapsed since Jeremiah Horrox, a native of Toxteth Park, and then only twenty years of age, observed the first transit of Venus across the Sun. His high attainments at that early period, in astronomy and pure mathematics, have been the admiration of succeeding men of science. His reputation may be said to have reached his native country from the continent, by the publication of his treatise *Venus in sole visa*, at Dantzic; and it is only of late years that Professor Rigaud and Mr. Whatton have laboured successfully to do justice to his memory."

The fame of Horrox has also been disseminated through the instrumentality of the press, letters having occasionally appeared, complaining that no record of his discoveries has been published in our native tongue, and commending the subject to the attention of those competent to deal with it. One of these, taken from the columns of a newspaper, was, a few months ago, enclosed to me by a friend, in which the writer thus alludes to the remarks of Professor Rigaud already quoted: "A life of Horrox is much wanted. Very little is known indeed of his daily work, but that little is such as to create a desire of knowing as much about him as possible. The particulars gathered up by Mr. Whatton will, I trust, be heard of, and make us better acquainted with one whom Sir J. Herschel justly calls 'the pride and boast of

British astronomy.' And surely the *Venus in sole visa* ought to have an English edition, for if, as Grant remarks, 'it does not redound to the credit of England that this exquisite relic of one of her most gifted sons should have been allowed to see the light in a foreign land,' neither does it evince a due regard for the labors of scientific men that this famous dissertation has yet to be published in our own country. I should be very much obliged for any information of the Whatton papers." Upon receipt of this extract, I searched for anything in my possession that might be available, and found sundry memoranda, and some interesting letters from Mr. Rigaud, the perusal of which led me to prosecute the inquiry until I was enabled to carry out, in some degree, the original design, by preparing a Memoir of the life of Horrox, and a translation of his discourse upon the transit of Venus.

It is felt that this little work is a very imperfect

substitute for what might have been achieved by abler hands; but being in possession of the details of Horrox's personal history, I should scarcely have been justified in withholding them, as it is a hopeless task for a stranger, on the spur of the moment, to attempt to look for such particulars as may be collected from a lengthened course of general reading. My aim has been to shew the value of his labors, and to fix the place they occupy in the history of science; and also to make his merits more widely known than they are at present, in order that he may enjoy in the estimation of the public, the rank which he already holds in the opinion of the learned. Accordingly, such letters and quotations as were written in Latin are here given in English. This will not occasion any confusion, as those which are translations may be distinguished at a glance from others which have been merely copied.

It will be observed that the name of Horrox is

sometimes spelt *Horrocks*. I have carefully examined which orthography is the more correct, and have adopted the former, as the name is so entered upon the College Register, and was always so written by Crabtree and Wallis. Grant and some recent authors use the latter method. The difference is of no importance, and it is only noticed here by way of explanation.

In the translation of the *Venus*, I have endeavoured to adhere closely to the original, and have taken the text of Hevelius as a basis, merely correcting the punctuation from the Greenwich manuscript where it was necessary to do so, and altering the arrangement of the sentences where the difference of language required it. The Dantzic edition is accompanied by voluminous notes which are appended to the end of each chapter, and at first I thought of giving them precisely in the order in which they stand. Afterwards it occurred to me that it would be better to print Horrox's dissertation entire, and to collect the notes together, and put them at the end by themselves, so as to present a clearer view of the treatise, without having the attention continually called off, sometimes indeed when there is no difficulty that needs to be explained At length, however, I decided to omit them altogether, as they contain nothing of importance connected with Horrox's personal history, and are full of error upon those points which they were designed to elucidate. The mistake that Hevelius has made in his statement of the parallactic angle is an instance of this, and has given rise to many faulty corrections in his comment. Flamsteed noticed it, and did not consider his remarks a very valuable appendage; for in a letter to Collins, he says: "Having well perused the *Venus in sole visa*, I know not what can be added; the notes of Hevelius I find generally useless, and those on the 6th chapter absolutely false." The side-notes which are found in the printed edition have also been excluded, as it is certain that they are not authentic. These accretions being removed, the tract appears in the same form, though not in the same dress, as that which it had when it came from the pen of its author; and the reader is enabled to peruse it without distraction, and to arrive at an independent opinion of its merits.

In writing what follows, I have consulted Ferguson, Delambre, Montucla, Grant's Treatise upon Physical Astronomy, and the suggestions of Professor Rigaud contained in the manuscripts in my possession. The correspondence between Huygens and Hevelius is taken from Huygens' papers preserved in the public library at Leyden. No doubt there is abundant room for criticism; but it may be pleaded that the task was wholly unsought, having devolved on me from circumstances over which I had no control, but from the obligation of which it would have been unworthy to retreat. Should these pages be deemed insufficient for the purpose which has been announced, I can only say that I shall be much gratified if some one, more competent than myself to do justice to the memory of Horrox, will make use of the material, here gathered together, to produce a better work. And I may add, as a further extenuation, that they have been penned in such brief intervals of leisure, during the last few months, as remained over and above the discharge of more important duties; so that I may fairly take refuge in Horrox's own words, "Ad majora avocatus, quæ ob hæc parerga negligi non decuit."

39, Weymouth Street, Portland Place,
July 26th, 1859.

MEMOIR OF JEREMIAH HORROX.

We are familiar with the names of some writers who have contributed scarcely anything of real value to the literature of their country; whilst we are ignorant of the worth of many others who occupy a distinguished position in the commonwealth of science. Thus few persons have heard of Jeremiah Horrox, although his merits as an astronomer have been acknowledged by the most eminent scientific men who have succeeded him. But he lived in obscurity, and died young. He was not permitted by an all-wise Providence to carry on his investigations for more than a few short years. He did not even enjoy the satisfaction of publishing his own discoveries. He was cut off in the midst of usefulness, and others have entered into his labors. Hence he is comparatively unknown. Happily his performances, as a skilful pioneer for the advancement of knowledge, are well authenticated, and are of sufficient importance to make his name illustrious. He paved the way for some of the most brilliant triumphs of the human intellect. Learned men have freely acknowledged this; and, in tender regard for the memory of one who expired whilst full of hope and promise, have constituted themselves the trustees of his reputation, and set their seal to his ability and worth. It is thought, therefore, that the details of his history may not be unacceptable, especially as his valuable services are now about to be recognised by a monument raised by subscription; and that the disinterested efforts of this young philosopher in search of truth cannot fail to enlist the sympathy and admiration of all who are made acquainted with them.

He was born at Toxteth Park, near Liverpool, in the year 1619. Little is known as to the position and circumstances of his family; but in the scanty notices of him that remain, he is generally spoken of as a person of humble origin. It seems probable, however, from his having been classically educated, and destined for one of the learned professions, that this representation is rather overdrawn, and that the Horrox family were not so obscure as they have been described. Liverpool was not then a seat of industry, enterprise, and intelligence, but a place of comparative insignificance; and Toxteth, far from being a wealthy and elegant suburb, was only a little village about three miles distant from it in the County Palatine of Lancaster. It is therefore extremely unlikely that he could have received any considerable advantages in his native place; and in those days, on account of the expenses of travelling and residence, it was not usual for a young man entirely without means to be sent to the ancient seats of learning. Hence we are led to conclude, either that his parents were in easy circumstances, and able to value the benefits of a liberal education, or that the genius of young Horrox attracted the attention, and secured the patronage of some person of distinction. Upon this and other points connected with his opening history, it is to be regretted that we possess so little information; for the auspices under which life commences, and the incidents of childhood, not unfrequently form an interesting page in the biography of great men. The school campaign, with its successes and failures, its schemes, friendships and amusements affords ample scope for the display of a boy's taste, talent, and disposition, and gives some indication of what may be expected from him in after life. Thus Isaac Newton, withdrawing from the noisy playground, spent his leisure hours in the construction of water-clocks, and other mechanical contrivances; Halley set up a sun-dial, and had observed the variation of the needle before he left school; Watt took an early pleasure in the manual exercises of his trade; James Ferguson made a watch of wood-work when quite a boy; and it is reasonable to suppose that Horrox in like manner shewed a partiality for the pursuits in which he afterwards distinguished himself. In those days lads of more than ordinary promise were admitted to the University much younger than they are at present, especially if introduced by an influential patron; hence we are not surprised to find that as soon as Horrox had received the rudiments of education at Toxteth, he was entered at Emmanuel College, Cambridge, before he had attained his fourteenth year. The following is a copy of the Register:

"Jeremiah Horrox. Born at Toxteth, Lancashire.
Entered Sizar, 18 May, 1632."

His having been placed on the college foundation, tends to confirm the surmise that his parents were not affluent, and that his advantages had hitherto been limited. But we know from the history of others who have attained to eminence in the several departments of learning, that the aspirations of genius cannot be wholly crushed by poverty, but that it will rise superior to circumstances, as surely as a blade of grass breaks through a clod of earth, and points its spire to the heavens. Horrox hailed with delight his removal from the village school to a seminary abounding with the means of intellectual improvement, and resolved to make the most of his opportunities. Having read the few subjects which were then included in an academical education, he explored the wide field of classical literature, readily yielding to its allurements, and regarding them as more than a compensation for any amount of labor. He particularly cultivated the best Latin authors, in order to become familiar with a language which was then the only medium of communication amongst the learned. In this way he acquired a large store of general knowledge, and was enabled to gratify his taste for any favourite pursuit. In a word, he drank deeply at the Castalian fount, and by his industry repaid the effort that had been made to send him to Cambridge.

But whilst he was fully capable of appreciating the advantages of an University, he did not remain at college longer than was absolutely necessary, being desirous of preparing for the work of the ministry, which he had adopted as the profession of his choice. Some doubt has been entertained as to whether he was ever admitted into Holy Orders. Young men are now required to be twenty-three years of age before they can be ordained, whereas he was not more than twenty. This objection might easily be answered by the fact that two centuries ago the question of age was not so strictly attended to, the Bishop exercising a discretionary power. But fortunately we are able to place the matter beyond conjecture; for in a treatise by John Gadbury, the compiler of almanacks, there is mention of

"Ephemerides of the planetary motions, eclipses, conjunctions, and aspects for fifty years to come, calculated from the British tables, composed first by the *Reverend* Mr. Horrox, and first published by Jeremy Shakerley."

He commenced his ministerial labours in his native county, being ordained to the curacy of Hoole, in Lancashire. This place formerly consisted of a narrow strip of land, having a large extent of moss on the east and west, the waters of Martin-Mere and the Douglas on the south, and the overflow of the Ribble on the north. It was therefore almost an island; and though doubtless an open situation for an astronomer, it could not have been a very agreeable residence. This once desolate spot is now a thriving township containing about a thousand inhabitants. The hand-loom and power-loom furnish their chief employment, though much of the land has been reclaimed, and is under tillage. The Parish Church, which was erected in the fifteenth century, is dedicated to Saint Michael, and consists of a plain brick nave without side-aisles, a chancel, and a stone tower supported by four pillars. There has long been an endowment for educational purposes, and about eight years ago a good national school and school-house were built after a plan by the government architect, at a cost of £600. Mr. Horrox's first letter from Hoole is dated June 1639, and he continued to reside there for some little time. There is no local record of his official connexion with the place, as it was not then constituted an ecclesiastical district, being merely a chapel of ease to the mother church of Croston, the register of which is comparatively modern; but that he was curate of the parish is a matter of history, for to omit the testimony of other writers, we may mention that Costard, an eminent astronomer who lived at the beginning of the last century, designates him as ' a young clergyman of Hoole, near Preston." There is reason to believe that, besides his ministerial avocations, he was in some way engaged in tuition, as he speaks of his "daily harassing duties" during the time he resided there.

It was whilst he was at the University that he first turned his attention to the study of astronomy. With a love of the sublime, and naturally fond of speculation, in the contemplation of the works of God he found a pursuit at once congenial to his taste, and calculated to bring into active exercise the highest powers of his mind. It did not satisfy him to look with an untutored eye upon the sun, the moon, and the stars shining in the firmament of heaven; he desired to learn something of their magnitudes, their distances, the periods in which they perform their revolutions, and the laws by which they are governed. "It seemed to me," he says, "that nothing could be more noble than to contemplate the manifold wisdom of my Creator, as displayed amidst such glorious works; nothing more delightful than to view them no longer with the gaze of vulgar admiration, but with a desire to know their causes, and to feed upon their beauty by a more careful examination of their mechanism." Animated with these convictions, he prepared to enter upon the study of astronomy by first cultivating with the utmost patience the aptitude for mathematics which he had evinced from his youth. But he had to work without assistance; for at that time, no branch either of mathematical or physical science was taught at Cambridge. In this respect she was considerably behind her sister University. Many scientific men had already emanated from the cloisters of Oxford. Bacon, Sacrobosco, and Greathead, were educated there. In short, the renown which Cambridge has acquired, and now enjoys in this kind of learning, is of a comparatively recent date. Certainly she had no school for science before the commencement of the seventeenth century. This was owing to the endowments of Oxford being older and richer, and to its collegiate system being earlier established. Thus he had no professional instruction; he could not obtain in the University the books he required; nor was there any one capable of advising him as to which it was most desirable for him to procure. This was particularly the case in reference to astronomy, which had scarcely yet taken root in our land. Its votaries had no measure of experience to consult, no body of doctrine to quote. Not a single public observatory had been erected either in England or France, nor indeed had astronomical observation as yet become fairly organized. The difficulty there was in obtaining works on physical science, may be illustrated by the following circumstance. Some time ago Mr. de Morgan met with a book which had formerly belonged to Horrox, and upon examining it, he found that it contained a written catalogue of the library which, at some period of his life, he seems to have possessed :—

Albategnius.
Alfraganus.
J. Capitolinus.
Clavii Apolog. Cal. Rom.
Clavii Comm. in Sacroboscum.
Copernici Revolunitiones.
Cleomedes.
Julius Firmicus.
Gassendi Exerc. Epist in Phil. Fluddanam.
Gemmæ Frisii Radius Astronomicus.
Cornelii Gemmæ Cosmocritice.
Herodoti Historia.
J. Kepleri Astron. Optica.
—— Epit. Astron. Copern.
—— Com. de motu Martis.
—— Tabulæ Rudolphinæ.
Lansbergii Progymn. de motu solis.
Longomontani Astron. Danica.
Magini Secunda Mobilia.
Mercatoris Chronologia.
Plinii Hist. Naturalis.
Ptolemæi Magnum Opus.
Regiomontani Epitome.
—— Torquetum.
—— Observata.
Rheinoldi Tab. Prutenicæ.
—— Com. in Theor. Purbachii.
Theonis Comm. in Ptolom.
Tyc. Brahæi Progymnasmata.
—— Epist. Astron.
Waltheri Observata.

Now it is very remarkable that, so far as we can ascertain, not one of these books had been published in our own country. The above interesting relic was sent to the authorities of Trinity College, Cambridge, with a request that it might be carefully preserved. The student of to-day can hardly enter into the feelings of a young man thirsting for knowledge, and circumstanced in the manner just described. Not to mention the public lectures, libraries, associations, and other advantages which belong to an University, every department of knowledge is represented to the general reader by so great an abundance of literature, that the only difficulty is to make the best selection. Elementary treatises are now published at prices which place them within the reach of the poorest scholar. And after all, books are the best teachers. The minds of many who have immortalized themselves and reflected honour upon their country, have been formed without any other assistance. But in the seventeenth century books were scarce and dear. We conclude therefore that there are no such drawbacks to be experienced now, as those which oppressed the student in science before the days of popular literature. Horrox labored under the greatest disadvantages, and hence he has all the more merit. He toiled up the sides of Parnassus without friendly assistance or encouragement. He meditated alone upon the abstruse subjects of philosophical enquiry. Having procured such treatises as he could afford to purchase, he qualified himself for the successful pursuit of the sublime science with which his name will ever be associated. But he has recorded his troubles in touching language :—

"There were many hindrances. The abstruse nature of the study, my inexperience, and want of means dispirited me. I was much pained not to have any one to whom I could look for guidance, or indeed for the sympathy of companionship in my endeavours, and I was assailed by the langour and weariness which are inseparable from every great undertaking. What then was to be done? I could not make the pursuit an easy one, much less increase my fortune, and least of all, imbue others with a love for astronomy; and yet to complain of philosophy on account of its difficulties would be foolish and unworthy. I determined therefore that the tediousness of study should be overcome by industry; my poverty (failing a better method) by patience; and that instead of a master I would use astronomical books. Armed with these weapons I would contend successfully; and having heard of others acquiring knowledge without greater help, I would blush that any one should be able to do more than I, always remembering that word of Virgil's

"'Totidem nobis animæque manusque.'"

Although astronomy had not taken firm root in our land before the time of Horrox, elsewhere it had considerably revived. Its cultivation in Europe was the commencement of a new æra. For the first two hundred years after its introduction upon the continent, little ground was gained; but subsequently men of genius and strength arose, who effectually exposed the absurd hypotheses then in vogue, put the science upon a right basis, and by delivering it from the trammels of superstition, led the way in a career of perpetual improvement. Thus Copernicus had re-established the old Pythagorean doctrine which places the sun in the centre of the system. This at once simplified all the planetary movements. The apparent revolution of the heavens was explained by the diurnal rotation of the earth. Tycho Brahé had enriched the science by a series of accurate observations. He had detected the lunar inequality, known as the *variation*; he had proved the path of the comet of 1577 to run out beyond the moon's orbit; and he had prepared, as the most valuable product of his labors, a catalogue of 777 of the fixed stars. Kepler had explained the laws of celestial motion. He had discovered that the planets move in elliptical orbits, with the sun in the lower focus; that the radius-vector describes equal areas in equal times; and that the squares of the periodic times of any two planets are to each other as the cubes of their mean distances from the sun. He had also some knowledge of the laws of gravitation. Galileo had greatly extended the limits of astronomical vision. Having heard that, by a combination of lenses, objects might be made to appear nearer to the eye, he ascertained the truth of the report; and improved the invention so much, that he was soon able to explore the heavens with his telescope, and to reveal new wonders to mankind. Milton alludes to his discovery of the inequalities of the moon's surface:

"The moon whose orb,
Through optic-glass, the Tuscan artist views
At evening, from the top of Fesolé
Or in Valdarno, to descry new lands,
Rivers, or mountains on her spotty globe."

Besides this he had detected the phases of the planet Venus, the four satellites of Jupiter, the spots on the Sun, Saturn's ring, and a multitude of stars too small to be seen with the naked eye. Thus by the genius of a few great men, the science was completely reconstructed, and enriched with much valuable learning. Its advancement was also hastened by the preparation of tables for facilitating the long and tedious calculations inseparable from astronomical pursuits. But improvement is unsteady in every department of human industry. Like the motion of the heavenly bodies, it is at one time accelerated, and at another retarded. An apostle or reformer suddenly appears, and promotes the welfare of his fellow men by rectifying abuses, and by bringing to light important truths. After he has delivered his message a calm ensues which lasts until another master-spirit arises. It is so in trade, politics, literature, and science; and it is wisely ordered that time should be allowed for testing by experiment the principles that have been broached. The three astronomers last named were contemporaries; and their departure was followed by a period of comparative inactivity. This was however very soon relieved by the appearance of Horrox, upon whom their mantle may be said to have fallen. But he did not take up the prophetic strain from the point where they had left it; he did not see the writings of his famous predecessors until after he had labored at astronomy for some time; he had to work out the grammar of the science for himself; to toil over ground that had already been surveyed; and being without friendly assistance, his worst fears of going astray for want of an able adviser were unfortunately realized. Happening to meet with a treatise by D. H. Gellibrand, a professor of astronomy, in London, in which the works of Lansberg were spoken of with unqualified praise, it occurred to him that it might be advantageous to possess them; and after some difficulty, he succeeded in obtaining the *Uranometriam*, the *Tabulas Perpetuas*, and the *Progymnasmata de motu Solis*. Pleased with the acquisition, he was induced to neglect the more valuable works of Tycho and Kepler, and to employ himself in computing Ephemerides from the tables of the

Flemish mathematician, not suspecting the speciousness of the titles which he prefixes to his calculations; but after a considerable time spent in this manner, he began to make his own observations, using these Ephemerides to point out the situations of the planets, and hence determining when their conjunctions, their appulses to the fixed stars, and other remarkable phenomena were to be expected.

In the year 1636, he made the acquaintance of William Crabtree, a draper, residing at Broughton, near Manchester, who had long been devoted to the study of astronomy; and a correspondence was at once commenced between them upon the various subjects connected with their favourite pursuit. This intercourse was the signal for increased assiduity on the part of both, and proved in one respect particularly useful to Horrox—it opened his eyes to the imperfection of Lansberg's tables. Hitherto, upon noticing a disagreement between them and his own observations, he had supposed the error was attributable to himself; and although the same result invariably followed after repeated trial, and there appeared to be no way of removing the discrepancy, rather than doubt the accuracy of one for whom he entertained so high an opinion, he continued equally self-suspicious, and was almost tempted to despair of success. But upon comparing notes with Crabtree, and perceiving that their observations entirely coincided, he called the attention of that gentleman to the circumstance, and was by him advised for the future to put less faith in the dictates of Lansberg. This led to a more rigorous examination, both of the tables, and also of the principles upon which they were based; and it soon became evident that much of what was put forth as truth was incapable of demonstration.

Emancipated from this tyranny of error, Horrox gathered fresh courage to proceed; he strove to redeem the time he had lost by redoubling his exertions; and afraid of being again misled by the misrepresentations of others, he learned to place more dependence upon his own judgment. At the same time he determined to avail himself of whatever aids and appliances he could obtain: new books and instruments were procured; and instead of seeking seclusion as before, he verified his operations by a regular correspondence with Crabtree. Besides this agreeable intercourse, the two friends presently became known to Dr. Samuel Foster, the Prælector of Gresham College, an able ally, whom they occasionally consulted.

The removal of a false impression, such as the one now described, if it does not give an actual impulse to the mind, at all events restores its wasted powers, and turns them to the best account. The clouds being dissipated, a new light breaks in, by which we can review the experience of the past, ascertain the strength of our present position, and lay down fresh plans for the future. Having escaped from the empiricism by which his expanding genius had so long been circumscribed, Horrox sought out the writings of Kepler, which Lansberg had stigmatized as "falsa et erronea, imo absurda, et inter se pugnantia." He instantly perceived their value. He found that instead of being composed of fanciful speculation, or arbitrary assertion, as he had been led to believe, they contained discoveries of such importance as to constitute a new era in the history of astronomy; and he received with transport the elucidation of general laws which were evidently the conclusions of a patient and legitimate induction. He also fully appreciated the merits of the Rudolphine tables, and considered them incomparably superior to those of Lansberg, as the hypotheses were well established, and reconcilable with one another. To amend these tables was now his chief desire. It occurred to him that they might be improved by changing some of the numbers, but retaining the hypotheses; and that he would be abundantly repaid for this arduous undertaking by the opportunity it would afford for deducing general principles, and especially for verifying Kepler's laws. Accordingly he applied himself to this task with unwearied diligence; and by making frequent observations, and altering the numbers to suit them where it was necessary, he brought the tables to a surprising degree of accuracy, and in doing so, materially added to his information. Speaking of the gratification he derived from the writings of Tycho and Kepler, and the incentive they were to renewed application, he says: "It was a pleasure to me to meditate upon the fame of these great masters of science, and to emulate them in my aspirations"; and accordingly we find that whilst he fully recognized the merits of the illustrious Dane as a skilful observer, his sagacious intellect clearly apprehended the truth of Kepler's doctrines, the universal acceptation of which he sought to promote.

The first efforts of Horrox's pen were directed towards the preparation of a treatise, the object of which was to refute Lansberg's theories, and to establish a more correct system of planetary distribution. He thought it important to the interests of science that the false hypotheses which then prevailed should be thoroughly exposed, and a misapplication of time and talent prevented for the future; and he wrote several learned dissertations, some of which were re-cast from beginning to end as often as it appeared to their author that they might be improved by a different mode of treatment. To specify a few of these, we may mention that at the close of the year 1637 he commenced a treatise entitled "*Jeremiæ Horroccii Anti-Lansbergianus, sive disputationes in astronomiam P. Lansbergii, quibus perspicue demonstra-*

tur, hypotheses suas nec cœlo nec sibi consentire." Having completed upwards of four disputations, he changed his plan, and re-modelling the whole, entitled it "*Astronomiæ Lansbergianæ censura et cum Kepleriana comparatio.*" Of this he wrote three copies agreeing with each other as to their object and arguments, but differing in the mode of discussion, and in their respective lengths: of the first copy he only finished one chapter, of the second nearly four, and of the third upwards of five. This favourite tract appears again in another dress, being designated as "*Explicatio brevis et perspicua diagrammatis Hipparchi, et Lansbergii erroris,*" but it is in substance the same as the former ones.

He next wrote a treatise against Hortensius, a follower of Lansberg, who had attempted unwarrantably to depreciate the merits of Tycho; and here also he seems not to have grudged the labor of repeated efforts in order to produce an essay that should be perfectly conclusive. Thus we have firstly a paper inscribed "*Contra Hortensii præfationem, Lansbergii Commentationibus de motu Terræ præfixam.*" This was afterwards re-written and styled "*Anti-Lansbergianus, seu astronomiæ veræ vindiciæ. Pars prima in qua respondetur Martinii Hortensii cavillis adversus Tychonem.*" Its title was again changed to "*Dissertatio cum Martino Hortensio de astronomia Tychonica.*" It was next called "*Astronomiæ Tychonicæ apologia, adversus Hortensii cavillas.*" And lastly, "*Epilogus ad Martinum Hortensium, in quo cavillis adversus Tychonem respondetur.*"

There are also other tracts upon similar subjects; for example, the commencement of a work entitled "*Præludium Astronomicum,*" of which the first book only "*de motu solis*" was in hand, a chapter of it upon the sun's horizontal parallax being entirely finished; the beginning of another treatise inscribed "*Anti-Lansbergius sive astronomia vindicata*"; and part of another, in which it was proposed to institute a comparison between various hypotheses of the system of the universe, which is inscribed as "*Paris Astronomicus, seu Judicium de vera astronomia, quo trium astronomorum Kepleri, Longomontani, Lansbergii tabulæ astronomicæ, et hypotheses, seu tabularum fundamenta, rationibus physicis, demonstrationibus geometricis, et observationibus astronomicis recentibus et antiquis ad examen mathematicum revocantur.*" These treatises exhibit much foresight and learning, and were well calculated to effect the object for which they were prepared, namely, to explode false doctrines, and to demonstrate the only rational hypothesis of our system.

Horrox next made some considerable improvements in the lunar theory. It is generally acknowledged, and indeed Sir Isaac Newton expressly states, that this young philosopher was the first person who discovered the moon's motion to be in an ellipse about the earth, with the centre in the lower focus. This discovery was not merely an extended application of the doctrines of Kepler. That great man had proved the ellipticity of the orbit of Mars, the earth, and other of the heavenly bodies, and had endeavoured to explain its cause; but Horrox, in his speculation on the moon's motion, outstripped the discernment of Kepler, inasmuch as he correctly explained the physical cause of the curvilineal motion of the planets, and shewed that it arises from the joint action of two separate forces. This was a great step in the progress of celestial dynamics. He tells us that he had spent much time in meditating upon the principle in virtue of which the planets describe oval orbits, and that he thought he had at length hit upon the true theory. Kepler had supposed them to be whirled round by the action of magnetic fibres, by which, as he thought, a mutual influence was exercised similar to that of the poles of loadstones; but being unable to reconcile the rotation of the sphere upon its axis with this supposition, he had recourse to the singular idea of the exterior only of the planet being endued with rotatory motion. Horrox states at some length his objection to this hypothesis, and having mentioned difficulties which Kepler himself had not perceived, he proceeds thus: "To say, as he doth, 'Hæc contemporatio pertinet ad consilium creatoris,' which I understand to be, so is the will of God, if it had come sooner might have saved a labour of all troublesome inquirys, for it is most true that the will of God is the cause of all things, but resting in generalitys is the death of philosophy.

I must have another cause of that ovall figure, which it is most certain all the planets do affect. This will not satisfy me." He then gives his own views, and says that, as the laws of nature are everywhere the same, there can be no doubt that the true principle of the ellipse may be illustrated by means of movements upon the surface of the earth, as for example, the throwing of a stone into the air, the rotation of which does not impede its progress. In this analogy, to which he refers more than once, we have the true explanation of celestial motion, now understood to be the combined effect of projective and attractive forces. If a stone be thrown obliquely into the air, its movement is governed by the impulse imparted to it by the hand, together with the attractive power of the earth. In obedience to these two influences, instead of tending in its fall directly towards the centre, it preserves whilst descending the same angle at which it arose; and if its progress were not interrupted by the earth's surface, there is little doubt that it would revolve unceasingly in an elliptical orbit with the centre in the lower focus. Hence arises the general law.—When two spheres are mutually attracted, if not prevented by foreign influences, their straight paths are deflected into curves concave to each other, and corresponding with one of the sections of a cone, according to the velocity of the revolving body. Thus if a sphere were projected by an independent power, as the planets were when launched forth from the Creator's hand, it would move forward in a right line for ever, unless attracted from it by an extraneous force; for instance, the earth would preserve a perfectly straight course whilst permitted to do so, but coming within the sun's influence, it is induced to deviate from the direction originally impressed upon it. Now if the velocity with which the revolving body is impelled be equal to what it would acquire by falling through half the radius of a circle described from the centre of deflection, its orbit will be circular; but if it be less than that quantity, its path becomes elliptical. This law was subsequently expanded by Sir Isaac Newton into the great principle of gravitation. As is well known, he concluded that the power which causes a body to fall to the earth, is of the same nature as that which retains the planets in their orbits; and he pursued this discovery, until he finally evolved an expression to which the phenomena of all the celestial movements may be confidently referred. Whilst thus engaged, he derived important assistance from the writings of Horrox, who, by his sagacious application of projectile to celestial motion, has gained a distinguished place amongst those whose labors have contributed to the establishment of the true system of the universe.

Having ascertained the ellipticity of the moon's orbit, and assigned its cause, he proceeded to examine the various inequalities which render the exact computation of her elements so difficult. If she were not subject to any foreign influence, the quantity of her ellipsis, the periods of her revolutions, and other particulars would always be the same; but as she is attracted by the sun as well as by the earth, the figure of her orbit is altered, and irregularities are occasioned which require to be corrected, in order that her theory may be satisfactorily developed. Horrox's enquiries led him to a distinct knowledge of the motion of the lunar apsides. He found that the longer axis of the ellipse, or that imaginary line which joins the apogee and perigee, moves slowly round the centre of the earth in the same direction as the moon revolves; and this change of position, which has since been ascertained to amount to rather more than three degrees for each of her sidereal revolutions, he rightly attributed to the perturbative influence of the sun. The beautiful experiment by which he illustrates this phenomenon shews not only that he was perfectly aware that an orbit might be formed by a central force, but also that within certain limits the heavenly bodies exercise a disturbing power upon each other. Crabtree had asked to be favoured with suggestions respecting the motion of the aphelion of a planet. In reply, Horrox, always adhering to his conviction of the harmony of nature and the possibility of exemplifying celestial movements by those which are common upon the earth, supposes a ball to be suspended by a long cord made fast to a hook in the ceiling. Now if the ball be drawn from the perpendicular, and then suddenly released, it oscillates for a while, with a speed which increases as the centre is approached, and diminishes when that point has been passed. But, if after having been withdrawn from the vertical, a tangential impulse be imparted, the ball will describe an ellipse; and what is particularly to be observed, the major axis will be seen slowly to advance in the same direction with the ball, performing, in course of time, a complete revolution. This illustrates the movement of the apsides of the lunar orbit; though in order to represent nature more correctly, the centre of force should be in the focus of ellipse, whereas in the experiment it is in the centre. Horrox perceived this defect in the illustration, and removed it by supposing a slight breeze to blow continually in the direction of the major axis, by which the relative situation of the point at rest would be changed. This ingenious experiment has been erroneously ascribed to Hooke, who reproduced it at a meeting of the Royal Society; but it was recorded as Horrox's invention more than five-and-twenty years before the idea was communicated to that learned assembly: and as the doctrines exemplified are of such importance, and were never before suggested by any astronomer, it is very fitting that he should have the credit of their discovery, and that the time when they were first brought to light should be correctly stated.

The principal irregularity affecting the place of the moon in her orbit, next to the equation of the centre, is usual called the *evection*, the existence of which was known to the astronomers of Greece. Its effect is to diminish the equation of the centre when the line of the apsides lies in syzigy, and to increase it when it lies in the quadratures; and it was explained by Horrox, as depending upon the libratory motion of the apsides, and the change which takes place in the eccentricity of the lunar orbit. This conclusion he arrived at from his own observation before he was twenty years of age.

He also determined the value of the *annual equation*, an inequality arising from the sun's perturbative influence, and which under ordinary conditions, is as the cube of his distance from the earth. It varies according to the position of the latter planet in its orbit as it approaches to, or recedes from its primary. It was noticed both by Tycho and Kepler, but neither of them assigned its quantity. Horrox stated its maximum value to be 11′ 16″ which is within four seconds of what it has since been proved to be, by the most accurate observations.

These improvements in the lunar theory, and the various doctrines which he has illustrated in connection with it, are alone sufficient to secure for him a lasting reputation. Perhaps he is more generally known by his other writings; but this is the subject in which his sagacity is the most conspicuous, and with which his name is the most honorably associated. Its accurate development has from time to time occupied the attention of the ablest astronomers; but it is not too much to say that his discoveries eclipsed the efforts of all his predecessors, and have been the foundation of the advancement towards perfection which has been made in modern times. His views were gradually unfolded in his letters to Crabtree, and are partly embodied in a systematic treatise, entitled "*Novæ Theoriæ lunaris, a Jeremiâ Horroccio primum adinventæ, et postea in emendatiorem formam redactæ, ex epistolis socii ipsius Gulielmi Crabtrei, ad eruditissimum virum Gulielmum Gascoignium scriptis, explicatio.*"

Another instance of his sagacity consists in his detection of the inequality in the mean motions of Jupiter and Saturn. This phenomenon results from the tangential impulse which is exercised to a remarkable degree by these two planets upon each other. It is a law of celestial mechanics that action and reaction are equal and in contrary directions, precisely as they are in reference to terrestrial bodies. If an anvil be struck, the reaction of the hammer is as great as the force communicated by the blow; and in like manner, one planet cannot impart momentum to another without subjecting itself to a corresponding influence. Consequently if the relative positions of Jupiter and Saturn in their orbits are such that the motion of one is accelerated, that of the other will necessarily be retarded; and a want of uniformity arises which in the instance before us is very important, on account of the extent to which it accumulates. Thus about the time that Horrox lived, and for a hundred and fifty years before, the mean motion of Jupiter was constantly increasing, and that of Saturn slackening; so that, upon examining the Rudolphine tables, he found that the calculated places of these planets did not agree with their true situations. Accordingly he suggested that the motion of Jupiter might be corrected by adding 1° 30′ to the aphelion, and 2′ to the mean longitude; and he estimated the quantity of acceleration at 1′ in ten years, which very nearly corresponds with the increment actually given to the mean longitude of Jupiter in each successive period of ten years during the first half of the seventeenth century. He also writes concerning the mean motion of Saturn, that sometimes it appears to be singularly retarded, and that in the time of Walther it was evidently slower than Kepler's calculations had made it; and he proposes to subtract 4′ from the planet's mean longitude at the beginning of the year 1600. He adds that the phenomenon would occasion him greater annoyance were it not for the consolation of his being in all probability the first person to discover it; and he requests Crabtree to make frequent observations for the purpose of finding out the correction to be applied to the Rudolphine tables. From various remarks which Horrox makes respecting the alteration in the lengths of the periods of these two planets, there is every reason to believe that he had conjectured the inequality of their mean motion to be periodic.

He bestowed considerable attention upon the nature and movements of comets. These bodies have at all times been regarded with great interest; not only by the ignorant, on account of their sudden and terrific appearance as the supposed harbingers of evil and the executioners of vengeance upon a guilty world, but equally by the philosopher who has labored to explain their extraordinary physical constitution, the irregularity of their movements, their apparent variations in size, and other peculiarities. They were for many ages believed to be only meteors confined within the orbit of the moon. Tycho was the first to refute this opinion by proving that they travel beyond Mercury or Venus. Horrox procured his treatise upon comets, and, without entirely adopting his suggestions, began to speculate upon the elements of their orbits. His reflections at different times shew how he advanced step by step in search of truth, his sagacious intellect laying hold

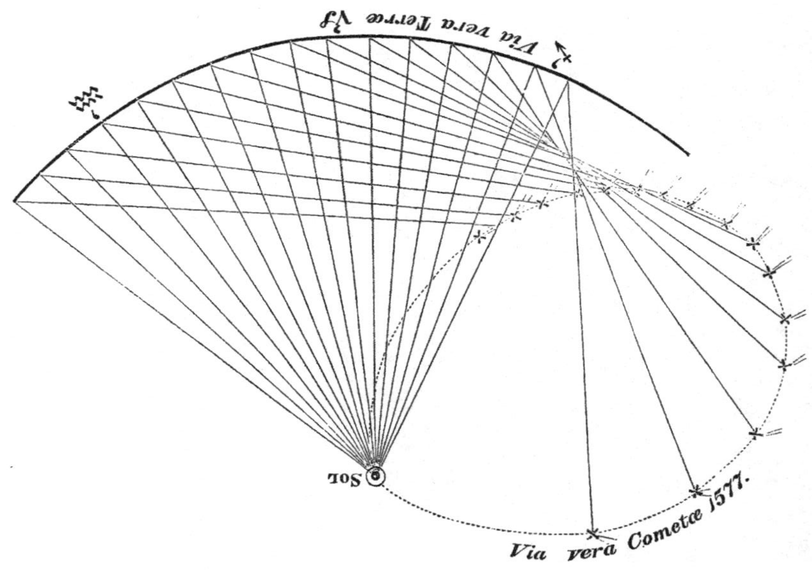
Sol
Via vera Terræ
Via vera Cometæ 1577.

of any outgrowth, and trying its strength to raise him from one firm footing to another. At first he conceived them to be projected from the body of the sun in straight lines, an opinion previously entertained by Kepler, and evidently suggested by the prodigious elongation of their orbits. He next assigned to them a velocity which diminishes as they recede from the sun, and increases as they return to it again. He then improved these conjectures by supposing their path to be curvilineal. Afterwards he says that they move "in an elliptical figure or near it," and illustrates this stage of his opinions by drawing a diagram for the comet of 1577 The orbit which he traces (see the figure) has an obtuse cusp at the sun, and could not really have been described; but it shews that he had arrived at the conclusion that comets revolve in curves returning into themselves. Wallis enclosed this diagram in a letter to the Royal Society, requesting that it might be carefully preserved, as it is in Horrox's own handwriting. Finally he determines that comets move "in elliptical orbits," being "carried round the sun" with a "velocity which is probably variable." This hypothesis has since been confirmed by a great number of observations, and is now generally received. It was however reserved for Newton fully to determine the elements of these bodies. He proved that any conic section may be described about the sun, consistently with the principle of gravitation; and also that these erratic bodies are subject to the general laws of planetary motion, notwithstanding the elongation of their orbits, and the unusual inclination of their planes to that of the ecliptic.

Horrox also commenced a series of observations on the tides. In his time very little was known as to their physical cause. As there are no tides in the Mediterranean, the ancients probably wrote of them from representation. Kepler explained their elevation more satisfactorily than any of his predecessors. Horrox proposed to investigate the subject thoroughly, and made various experiments for the purpose of ascertaining the extent of their rise and fall at different times, and at different places, their direction, and the influences to which various phenomena respecting them are to be attributed. After he had continued his labors for three months, he wrote to Crabtree, telling him that he had noticed many interesting particulars which had not then been remarked by any one, and that he hoped before long to arrive at some valuable conclusions respecting their nature and cause. Unfortunately we do not possess the result of his observations, no papers containing a systematic account of them having come down to our times. We must however allow him the credit of being the first person to undertake a regular course of tidal observations, for the purpose of philosophical investigation.

It is worthy of remark that he approved of, and frequently employed a decimal system of arithmetic. Since the commencement of the seventeenth century, great improvements, most of which are based upon the introduction of the decimal principle, have been made towards abridging the labour of calculation. This method was invented by one Simon Steven, a native of Bruges, in 1602, and it prepared the way for the discovery of logarithms by Sir John Napier, within twelve years afterwards. Horrox strongly recommends the adoption of a decimal notation, wherever it can be successfully applied ; and he expresses his opinion that it would have been better if the circle had been divided into 100 or 1000 parts, instead of 360. He says that such an arrangement would have been preferable to any other, and that the sexigesimal division is attended with many inconveniences. He also proposed to publish ephemerides in this form, in order that astronomers might have an opportunity of judging of its merits. Public attention in England has of late years been particularly directed to this subject, and much has been said and written to prove that the application of the decimal principle to our coinage would simplify the course of exchange, and make the reckoning of money more intelligible to every capacity; but, admitting that such an alteration can only be brought about by slow degrees, it is doubtful whether the efforts that have been made for its adoption, are at all commensurable with the advantages that would follow; and it does not lessen our appreciation of Horrox's acuteness, to reflect that he approved and employed a mode of calculation which has yet to be introduced into many departments of practical business.

Whilst he was studying the writings of Lansberg, he was led to conclude that there would be a transit of Venus in 1639. The calculations of the Flemish astronomer respecting the motions of this planet are for the most part very inaccurate. This obliged Horrox carefully to re-consider them, and in so doing he discovered to his great joy that the conjunction was to be expected. In order to satisfy himself thoroughly upon this interesting point, he consulted the Rudolphine tables, by which his anticipations were confirmed. Strange to say, it does not appear that Kepler had any idea that a transit would take place in 1639; for in a little work published at Leipsic in 1626, entitled "*Admonitiuncula ad Curiosos rerum Cælestium*," he says, that Venus will pass over the sun's disc in 1631, and not return thither again until 1761. According to Hevelius no transit was witnessed at the former date, and the inaccuracy of the announcement may be traced to the imperfect state of the Rudolphine tables. Kepler died about twelve months before the time at which it should have happened; but Gassendi sought for it at Paris, and although the sky was clear, and he watched during the greater part of three days, he did not see Venus in the body of the Sun. The consequences of this mistake might have been disastrous to the interests of science; for the assertion that there would be no transit until 1761, had the effect of preventing astronomers from looking out for that of 1639, which took place on the 24th of November (Julian style) as Horrox had calculated, and which but for his foresight would not have been observed. It may not be out of place to remark, that it is now ascertained that the periods between the transits of Venus are 8,235, 243, and 713 years; so that by adding any of these numbers to the date on which some previous one is known to have happened, the result gives the time when another may possibly occur. There will be two more transits of Venus, in the ascending node, during the present century, *viz.*, December 8th, 1874, and December 6th, 1882, the latter of which will be visible in this country. The observation is of considerable value, as it affords means for correcting the planet's elements, and for determining the sun's horizontal parallax.

As soon as Mr. Horrox had satisfied himself as to the time of the conjunction, he wrote to inform his friend Crabtree that it was to be expected, and requested that he would make what observation he could with his telescope, and especially that he would carefully examine the planet's diameter which, in his opinion, had been considerably overestimated. He also begged him, if time allowed, to communicate with Dr. Foster, as it was desirable that the conjunction should be observed in several places in order to prevent the possibility of failure in case the heavens should be overcast. His letter is dated, Hoole, October 26th, 1639, and he says—"My reason for now writing is to advise you of a remarkable conjunction of the Sun and Venus on the 24th of November, when there will be a transit. As such a thing has not happened for many years past, and will not occur again in this century, I earnestly entreat you to watch attentively with your telescope, in order to observe it as well as you can. Notice particularly the diameter of Venus, which is stated by Kepler to be 7′, and by Lansberg to be 11′, but which I believe to be scarcely greater than 1′. If this letter should arrive sufficiently early, I beg you will apprise Mr. Foster of the conjunction, as, in doing so, I am sure you would afford him the greatest pleasure. It is possible that in some places the sky may be cloudy, hence it is much to be desired that this remarkable phenomenon should be observed from different localities." He adds that according to the Keplerian tables the conjunction will be visible at Manchester at 8h. 8m. a.m., the latitude of the planet being 14′ 10″ south, but that, according to his own correction, it should be seen at 5h. 57m. p.m., with 10′ south latitude. But inasmuch as a slight change in Kepler's numbers would considerably alter the quantity of the planet's latitude, it would be desirable to watch during the whole day, and also on the preceding evening, and following morning, although he did not doubt but that the transit would take place on the 24th.

After having deliberated on the best method of making the observation, he determined to admit the sun's image into a dark room, through a telescope properly adjusted for the purpose, instead of receiving it through a hole in the shutter merely, as recommended by Kepler. He considered that by the latter method the delineation would not be so perfect, unless it were taken at a greater distance from the aperture than the narrowness of his apartment would allow; neither was it likely that the diameter of Venus would be so well defined: whereas his telescope, through which he had often observed the solar spots, would enable him to ascertain the diameter of the planet, and to divide the sun's limb with considerable accuracy. Accordingly, having described a circle of about six inches diameter upon a piece of paper (see the plate), he divided its circumference into 360 degrees, and its diameter into 120 equal parts. This diagram was, in his opinion, sufficiently large for all practical purposes, nor did he think it necessary to carry the subdivision further, as he could depend upon the judgment of his eye with as much confidence as upon any mechanical arrangement he could then contrive. When the proper time came, he adjusted his apparatus so that the image of the sun should be transmitted perpendicularly to the paper, and exactly fill the circle he had described. From his own calculations he had no reason to expect that the transit would take place, at the earliest, before three o'clock in the afternoon of the 24th; but as it appeared from the tables of others that it might occur somewhat sooner, in order to avoid the chance of disappointment, he began to observe about mid-day on the 23rd. Having continued to watch with unremitting care for upwards of four-and-twenty hours, excepting during certain intervals of the next day when, as he tells us, he was "called away by business of the highest importance, which could not with propriety be neglected," he was at length rewarded for his anxiety and trouble by seeing a large dark round spot enter upon the disc of light. This was beyond doubt the commencement of the transit, as the solar spots are very rarely spherical, and do not consist of matter so regularly disposed, nor so dense, especially about the edges, as the object which he observed. They are generally composed of an umbra, or dark space, which is surrounded by a fainter shade. Venus could not have presented this appearance, as her shadow would be of an equal intensity of darkness and of a circular shape. He therefore examined it attentively, and arrived at some important conclusions. With respect to the inclination, he found by means of a diameter of the circle set perpendicularly to the horizon, the plane of the circle being slightly sloped on account of the sun's altitude, that to all appearance in the dark chamber, the planet was wholly immersed by a quarter past three, at about 62° 30′, from the vertex on the right hand, and that this inclination continued constant until sunset. He also accurately measured the distance of the Sun's and Venus' centres at various times during the transit. And he confirmed his previous conjectures respecting the planet's diameter, inasmuch as it only exceeded a thirtieth part of the diameter of the sun by about one-fifth subdivision, so that the proportion between them would be as 30′ to 1′ 12″, or at least to 1′ 20″; and this was evident in every situation of Venus. Thus the observation was well executed, and the results in all respects such as he had anticipated. The inclination was the only point upon which he was not quite satisfied, as he was unable to estimate it with very great exactness on account of the rapidity of the planet's motion. Hevelius thinks he might have used with advantage the method employed in observing solar eclipses, by which the sun's image would have been prevented from going beyond the paper, the apparatus having an observatory circle and a small table fixed at the end of the telescope, so that the most rapid motion of the sun could not have disturbed the observation; but he forgets that even if Horrox had thought of such a plan, his means were probably too limited to allow of his procuring the apparatus. The transit was witnessed at Hoole, the little village before mentioned, of which he was the curate. Its latitude is stated to be 53° 35′, and its longitude about 22° 30′ from the Fortunate Islands, or 14° 15′ west of Uraniburg.

With reference to the important business owing to which, we have said, he was obliged to leave his telescope, Hevelius further tells us, that he would not have suffered his attention to have been withdrawn by any occupation whatever, which could have been undertaken at another time; but that he would have watched Venus more assiduously than he had observed Mercury on a previous occasion, and that he would never have moved his eye from the circle unless some one else had been ready to take his place. But Horrox's absence is fully justified by the fact that the business which called him away was the discharge of his ministerial duties. Little calculation is necessary to prove that the 24th of November 1639, old style, happened on a Sunday; and the hours when he was obliged to relinquish his occupation correspond with those at which probably he would be engaged in conducting divine service. The following extract in support of this opinion will be read with interest It is copied from one of Thomas Hearne's pocket books, and dated February 8th, 1723—" Mr. Horrox, a young man, minister of Hoole, a very poor pittance, within four miles of Preston, in Lancashire, was a prodigy for his skill in astronomy, and had he lived, in all probability, he would have proved the greatest man in the whole world in his profession. He had a very strange unaccountable genius, and he is mentioned with great honor by Hevelius upon account of his discovery of Venus in the Sun, upon a Sunday; but being called away to his devotions, and duty at church, he could not make such observations, as otherwise he would have done."

When Crabtree was informed of the expected transit, he prepared to observe it in the same manner as his friend. But he was not equally successful; for though he watched most attentively, the sky was so over-cast that the sun could not be seen. At about 3h. 55m. by the clock, the clouds suddenly cleared away, when to his delight he saw Venus fully entered upon the Sun's disc. Overcome with rapture, instead of improving the opportunity thus favorably presented to him, he stood gazing at the spectacle without using his apparatus, nor did he recover his self possession until the heavens were again obscured. This may provoke a smile from those who know not the overpowering emotion which attends success in a painful and laborious pursuit; but let them remember that such intervals of satisfaction are the only reward which the astronomer receives for his toils of mind and body, for his watchings by night and by day, and for his tedious calculations and patient study. Every inventor and discoverer has his moments of ecstacy. When Pythagoras had fairly demonstrated the great geometrical truth, that the square described on the hypothenuse of a right-angled triangle is equal to the squares constructed upon the other two sides, such was his exultation that he forthwith sacrificed a hundred oxen to the gods; Archimedes, having discovered a method of ascertaining the specific gravity of different bodies, was so overjoyed as to forget the proprieties of life. Thus Crabtree is not the only person who has lost his self-control in a moment of transport. Nor did he entirely fail to take notice of what he saw; for though he was unable accurately to measure either the distance of the centres, or the angle of inclination, he made a sketch from memory of the planet's relative situation, which corresponded with what Horrox had observed, and he estimated its diameter at $\frac{7}{200}$ of that of the sun. This observation was made at Broughton, near Manchester, where Crabtree resided, the latitude of which is 53° 24′, and the longitude 23° 15′. Horrox also apprised his brother Jonas of the coming transit; but the unpropitious state of the weather prevented him from profiting by the information. It is believed that this phenomenon was not seen by any one except the two friends; and although the observation was made by both under unfavorable circumstances, it has been of considerable advantage to the science of astronomy. Horrox determined the position of the nodes, and the elements of the planet Venus with greater accuracy than had hitherto been attained. He also found that the time of the conjunction was 5h. 55m., instead of 5h. 57m. as he had anticipated; that the planet's latitude was 8′ 31″ south, instead of 10′; he concluded that the nodes ought to be placed at 13° 22′ 45″ from Sagittarius and the Twins, rather than 13° 31′ 13″ where Kepler placed them; and that of all the tables then in use, the Rudolphine were the most exact.

It may not be out of place to insert here a letter from Crabtree to Gascoigne, an able mathematician, and the inventor of the micrometer, as it refers to the observation, and is otherwise interesting as shewing the friendship and esteem which the writer felt for Horrox. After discussing various theories respecting the spots on the sun, and giving his opinion upon some philosophical experiments, Crabtree says :—

" In the mean time let me encourage you to proceed in your noble optical speculations. I do believe there are as rare inventions as Galileo's telescope yet undiscovered. My living in a place void of apt materials for that purpose makes me almost ignorant in those secrets : only what I have from reason, or the reading of Kepler's *Astronomia Optica*, and Galileo. If you impart unto us any of your optical secrets, we shall be thankful and obliged to you, and ready to requite you in anything we can. It is true which you say, that I found Venus' diameter much less than any theory extant made it. Kepler came nearest, yet makes her diameter five times too much. Tycho, Lansberg, and the ancients about ten times greater than it should be. So also do they differ as widely in the time of the conjunction. By Lansberg the conjunction should have been 16h. 31m. before we observed it : by Tycho and Longomontanus 1 day 8h. 25m. before : by Kepler, who is still the nearest the truth, 9h. 46m. before. So that had not our own observations and study taught us a better theory than any of these, we had never attended at that time for that rare spectacle. You shall have the observation of it when we see you. The clouds deprived me of part of the observation, but my friend and second self Mr. Jeremiah Horrox, living near Preston, observed it clearly from the time of its coming into the sun, till the sun's setting; and both our observations agreed, both in the time, and diameter most precisely. If I can, I will bring him along with Mr. Townley and myself to see Yorkshire and you. You shall also have my observations of the sun's last eclipse here at Broughton, Mr. Horrox's between Liverpool and Preston, and Mr. Foster's in London. Lansberg on eclipses, especially the moon, comes often nearer the truth than Kepler, yet it is by packing together errors; his diameters of the sun and moon being false, and his variation of the shadow being quite repugnant to geometrical demonstration. His circular hypotheses, Mr. Horrox, before I could persuade him, assayed a long time with indefatigable pains and study to correct and amend; changing and turning them every way, still amazed and amused with those lofty titles of perpetuity and perfection so impudently imposed upon them; until we found, by comparing observations in several places of the orbes, that his hypotheses would never agree with the heavens for all times, as he confidently boasts; no, nor scarce for any one whole year together, alter the equal motion, prosthaphæresis, and eccentricity howsoever you will. Kepler's ecliptick is undoubtedly the way which the planets describe in their motions; and if you have read his commentary '*de motu Veneris*," and his '*Epitome Astronomiæ Copernicæ*,' I doubt not you will say his theory is the most rational, demonstrative, harmonious, simple, and natural that is yet thought of, or I suppose can be; all those superfluous fictions being rejected by him, which others are forced so absurdly to introduce; and although in some respects his tables be deficient, yet being once corrected by due observations, they hold true in the rest, which is that argument of truth which Lansberg's and all others want. Your conceit of turning the circle into 100,000,000 parts were an excellent one, if it had been set on foot when astronomy was first invented. Mr. Horrox and I have often conferred about it. But in respect that all astronomy is already in a quite different form, and the tediousness of reducing the tables of sines, tangents, and all other things we should have occasion to use into that form; as also the inconveniences which we foresaw would follow in the composing of the tables of celestial motions, together with the greatness of the innovation, deterred us from the conceit. Only we intend to use the centesmes, and millesmes of degrees, because of the ease in calculation. I have turned the Rudolphine tables into degrees and millesmes, and altered them into a far more concise, ready, and easy form, than they are done by Kepler. My occasions force me to put an abrupt end to my unpolished lines, and without more compliments, to tell you plainly, but sincerely, I am your loving friend, (though *de facie ignotus*) WILLIAM CRABTREE.

From my house in Broughton, near Manchester, this 7th of August 1640."—The superscription of the letter is "To his loving friend Mr. William Gascoigne, at his father's house, in or near Leeds, Yorkshire."

It appeared desirable to Horrox, for many reasons, that an account of the transit should be prepared for the press, and accordingly he wrote an elegant treatise entitled "*Venus in sole visa, seu tractatus Astronomicus, de nobilissima solis et Veneris conjunctione, Novembris die* 24 *Styl. Juliano* MDCXXXIX, *autore Jeremia Horroxio,*" detailing the history of the observation and its value to the interests of science. But not being versed in the mysteries of authorship, and wanting means, he was at a loss to know how to procure its publication. He therefore requested Crabtree to write to his bookseller who would probably be able to advise them in this matter. After a few letters had been interchanged without anything satisfactory being concluded, he determined to accept a long-standing invitation to visit his friend at Broughton, which would enable him to discuss the subject more freely, to confer upon different points connected with their astronomical pursuits, and more especially to give the right hand of fellowship to one for whom he had so high a regard. He had more than once before purposed spending a few days with him, but his intention had as often been frustrated by the unsettled state of his affairs. At length, in order to fix some definite time, he wrote a letter from Toxteth, dated 16th December 1640, in which he arranged his journey for the 4th of January, and told Crabtree that he might expect him on that day, "if nothing unforeseen should occur." This is the language of one who felt the uncertainty of all human affairs, and was accustomed to act as not knowing "what a day may bring forth." His purpose was never carried into effect. To the inexpressible grief of every true philosopher, his short but brilliant career was closed by death the day before he should have arrived at Broughton. He expired on the 3rd of January 1641, in the twenty-second year of his age. As the flower of the morning falls before the scythe, so was he cut off in the freshness and vigour of youth. But his death was timely. His work was done. He went to the grave in a full age.

Having seen the glory of God afar off, his spirit soared to the heaven of heavens to worship Him as the centre of light and power. It is to be regretted that the particulars of his decease are nowhere recorded, and that we are left to mere conjecture upon a point of so much interest; but there can be little doubt that whatever may have been the immediate cause, his incessant labours by night and by day materially contributed to hasten it. Crabtree felt his loss acutely. His rapid and comprehensive understanding had removed many a difficulty from the path of knowledge, his sympathy had lightened many a toil. On the back of the letter last-mentioned, which was found tied up with several others, was the following touching inscription in Crabtree's handwriting:—

"Letters of Mr. Jeremiah Horrox to me, of the years 1638, 1639, 1640, until his death on the morning of the 3rd of January, when he expired very suddenly, the day before he had proposed coming to me. Thus God puts an end to all worldly affairs! and I am, alas! bereaved of my dearest Horrox. Irreparable loss! Hence these tears!"

The banishment of Tycho was overruled to the

advancement of astronomy, for it was owing to this circumstance that Kepler obtained possession of his theories and observations, which he afterwards re-produced and improved with such advantage to the scientific world. But in the remote part of the country in which Horrox died, no one was found capable of appreciating the value of his papers; and consequently, instead of being carefully preserved and kept together until they could be revised with a view to publication, many were destroyed, and the rest were carried away to different places. Thus one portion of them, which had been hastily concealed on account of the troubles of the times, was discovered and committed to the flames by a company of soldiers who entered his father's house in search of plunder. Another portion was appropriated by his brother Jonas, who carried them over to Ireland, where he died far from home and friends, and the papers were never afterwards recovered. A third fell into the hands of Jeremiah Shakerley, and was made use of by him in the compilation of the British tables published in the year 1653. He subsequently went out to the East Indies;

but before his departure entrusted his literary effects to one Nathaniel Brooks, a London bookseller, in whose possession they remained until they were burnt in the great fire of September 1666. The only papers that escaped these disasters were found in the house of Mr. Crabtree, who, knowing their intrinsic merit, had claimed them on the ground of past association; and influenced by motives of affection and esteem for their author, had preserved them with the utmost fidelity. It is not known how long this gentleman survived his friend. There are a variety of statements upon this point; but the greater number of them lead us to believe that he followed him to the grave within a very few years. When his establishment at Broughton was broken up, and his library about to be sold, these manuscripts, including that of the Venus, were discovered by Dr. John Worthington, Fellow of Emmanuel College, Cambridge, a man of distinguished piety and learning, who had been contemporary with Mr. Horrox at the University. In a letter dated 28th of April 1659, addressed to Hartlib, who had asked to see the dissertation on the transit, he says:—

"I have, as you desire, sent you Mr. Horrox, his discourse called "*Venus in sole visa.*" Here are two copies of it, but neither writ to the end. I lent them some years since to a friend who promised out of both to make out one, and then to print it; but other business it seems would not permit him to go through with the work. In some other loose papers I perceive that the author began his tract again and again (so curious was he about it), but these seem to be his last, written with his own hand. He lived at Toxteth Park near Liverpool, in Lancashire, was some time of Emmanuel College, Cambridge, admitted the same year I was. These papers of his (with many others of astronomical observations) I found in the study of one Mr. Crabtree (a Lancashire man, and his great correspondent in these studies), and I bought them after his death. By sending to some friend about Liverpool or Toxteth, it may be known whether any of Mr. Horrox's kindred have any of his papers.

"Yours, &c.,

"J. WORTHINGTON."

Hartlib having obtained the manuscript of the transit did not return it as soon as was expected. This appears to have caused the doctor great anxiety, and some little annoyance; for the following year he wrote to desire that it might at once be transcribed and sent back, as he did not think there was another copy of it extant. He also says, lest he should be thought uncourteous, that, as intimated in his previous letter, it had been borrowed on a former occasion by a person who had professed a wish to publish it, a measure which he entirely approved; but he adds that "all who design good things do not persevere when it comes to a business of some labour." A singular fatality seems to have attended these papers, and to have fully justified the anxiety that their owner had expressed concerning them; for while they were in Hartlib's possession, his study was burnt down, and they were with difficulty saved from destruction.

In the year 1660, a copy of the "*Venus in sole visa*," possibly one of those which belonged to Dr. Worthington, came into the hands of Huygens, the Dutch astronomer, who having been asked by Hevelius whether there was anything new going on in the scientific world, said that he could supply him with a copy of Horrox's celebrated observation. Upon this announcement Hevelius promised that if he would transmit it by the first opportunity, it should be published with annotations under cover with his account of the transit of Mercury which was then nearly ready for the press. After some delay it was forwarded; and when Hevelius received it, he expressed his satisfaction that the two tracts were to be made into one volume, in an eloquent strain: "How greatly does my Mercury exult in the joyous prospect that he may shortly fold within his arms Horrox's long-looked for, and beloved Venus. He renders you unfeigned thanks that by your permission this much-desired union is about to be celebrated, and that the writer is able with your concurrence to introduce them both together to the public." The annotations that were appended are very voluminous, being of greater length than the treatise itself. They were evidently written under unfavorable circumstances. Their author was at the time overwhelmed with affliction, and it is clear that they were somewhat hastily drawn up; for besides that they contain errors which could not possibly have remained if proper time had been allowed for revision; the work was out of the printer's hands in about three months after Hevelius had received the manuscript of the Venus, and a copy of it sent to Huygens with an accompanying letter, dated May 1662, to this effect:—

"You have doubtless heard, much honored friend, of the severe domestic calamity by which I was prevented from more quickly fulfilling my promise; and I am sure you will not only readily excuse me, but sympathize with me in this trial, when you understand how grievous an affliction has befallen me. I have sent you by Dr. Peltrius my Mercury produced amidst great mental anxiety, together with Horrox's Venus, happily risen for the public good, whilst alas! my own beautiful Venus has set to my infinite sorrow! I pray you to consider them carefully, until I am able to send you something better. The learned world is particularly indebted to you for bringing Horrox's Venus to light, thus having cheerfully bestowed a gift so excellent and acceptable as to demand the thanks of the latest posterity. When you have read the book, I beg you will give me your opinion of its merits, which I shall esteem a great kindness, and in turn you will always find me desirous of serving you."

To which Huygens replied on the 25th of July 1662:

"Your most acceptable letter, and shortly afterwards the volume of the new observations reached me safely, and although I ought to have thanked you before for the valuable gift, I have been so hindered that I could not until now discharge this duty. The illustrious Bullialdus informed me of the great affliction you have sustained by the death of your dearest wife, on which account I feared that this little work, which was then in hand, would be delayed. But you have acted rightly in not suffering your private loss to become a public misfortune; for I cannot say how highly astronomy is indebted to you for so accurate a description of your beautiful observation. Posterity cannot adequately repay you with its thanks. Touching the posthumous work of Horrox now brought to light, it is more satisfactory that it should have been undertaken by you, than by me; especially as you have prepared an excellent and elegant edition, and increased its value by a commentary. Furthermore, as you ask me freely to give you my opinion of the several particulars treated in the book, I frankly confess that your new method of ascertaining the diameters of the planets by that of Mercury appears less certain to me than to you."

The manuscript which was sent to Hevelius by Huygens does not appear to have been returned to him, as it is not among his papers in the public library at Leyden.

It is remarkable that, in Hevelius' edition of the *Venus*, the name of the place where the observation was made is nowhere to be found. But this circumstance is not attended with any difficulty, as the transit is described to have been seen fifteen miles north of Liverpool, which exactly corresponds with the situation of Hoole. It is clear that Horrox was residing in this village at the time of the conjunction, as all his letters between the months of June 1639 and July 1640 are dated from thence; and moreover the name of the place is inserted in the catalogue of his observations. The work is now extremely scarce; there are probably not half a dozen copies of it in the kingdom.

In February 1663, the subject of Horrox's manuscripts was brought before the Royal Society, and after some discussion, two of its members were instructed to procure from Dr. Worthington any papers which he possessed, for the purpose of being revised and published at the society's expense. These were readily obtained, and were entrusted to Dr. Wallis, the learned professor of geometry at Oxford, who having been desired to peruse them, reported upon their merits to the society at considerable length. He said that he and his colleague Dr. Christopher Wren had attentively examined them, and that in their joint opinion, what was written in English, consisting merely of notes from memory and unconnected paragraphs produced at various times, was unsuitable for publication; but that they considered the Latin pieces to be extremely valuable, and well worthy of preservation. Wallis was hereupon requested to gratify the learned world by digesting and preparing such of the manuscripts as he approved, a task which he gladly undertook, and which he was admirably qualified to fulfil. The plan that he adopted was as follows: By judiciously arranging the various tracts and dissertations put into his hands by the society, including especially those against Lansberg and Hortensius, with others already mentioned, he compiled a perfect treatise, entitled "*Astronomia Kepleriana defensa et promota.*" This is divided into seven disputations, with an introduction instituting a comparison between the merits of Ptolemy, Copernicus, Lansberg, Kepler, and others. The first dissertation is upon various hypotheses, and the formation of tables of the heavenly motions; the second upon the fixed stars; the third upon the obliquity of the Zodiac; the fourth upon the semi-diameter of the sun; the fifth upon the diagram of Hipparchus; the sixth upon the movements of the stars; and the seventh contains an answer to the cavils of Hortensius against Tycho. Whilst the manuscripts were in course of preparation, several other papers and letters were discovered, which were likewise carefully collated, and printed by Wallis under the same cover as those just enumerated. They consist of extracts from Horrox's letters to his friend Crabtree upon different astronomical subjects, a catalogue of his observations, his new theory of the moon, together with Flamsteed's lunar numbers upon it, also Crabtree's observations at Broughton, and Flamsteed's treatise upon the inequality of the solar year. The Astronomer Royal himself explains the circumstances under which his essay and numbers were appended to this collection of Horrox's writings. In his Autobiography, published some time ago by the Admiralty, we read:—

"I made a journey into Lancashire, and called at Townley, to visit Mr. Christopher Townley, who happened to be then in London. But one of his domestics kindly received me, and shewed me his instruments, and how his micrometer was fitted to his tubes; and from this time forward we often conferred by letters. I procured Mr. Gascoigne's and Mr. Crabtree's papers from him, and Mr. Horrox's theory of the moon, to which he had begun to fit some numbers; but perfected none that I remember. About this time Mr. Horrox's remains and observations, having been collected by Dr. Wallis, were in the press. I found his theory (of which a correct copy had fallen into my hands) agree much better with my observations than any other. Hereupon I fitted numbers to it, which with an explanation of it were printed with his works. Mr. Collins advised me to print my discourse concerning the equation of natural days with them: which I consented to do; and sent it up to him for that purpose translated into Latin."

All these papers combine to form a quarto volume, which was published at the expense of the Royal Society. Wallis announced the completion of his work in a letter to that learned body dated September 21st, 1664, in which he informed them that he had compared the different copies with the originals, arranged the several subjects in their proper places, and prefixed to the whole an epistle dedicatory to their president, Lord Brounker. A vote of thanks was then passed to Dr. Wallis, and the printing of the book was next referred to the consideration of the council; but owing to the low state of the society's funds at this early period of its history, the volume was not issued until nearly eight years afterwards. Its publication is mentioned in a quaint letter from John Collins to Dr. Edward Bernard, written on—

"16 March, 16$\frac{70}{71}$. From my house next the three Crowns in Bloomsbury market." He says, "Dr. Wallis, his comment on the astronomicall remaines of Horrox, is to goe to the Presse here, and there is a new type provided for the same, the Doctor desired to revise it first, that he might adde a running title to the Topp, I sent it on this day three weekes by Dobbins, Moores coachman, giving notice to the Doctor thereof by the Post and since wrote to the Doctor, but receiving no answer am afeard the Doctor is by his disease incapacitated, or under some great affliction."

The book at length made its appearance, being entitled "*Jeremiæ Horroccii Angli Opera Posthuma; una cum Gul. Crabtrei observationibus cœlestibus; necnon Joh. Flamstedii de temporis æquatione diatriba numerisque lunaribus ad novum lunæ systema Horroccii,*" printed in London, 1672. In the years 1673 and 1678 it went through two fresh editions, but was so inaccurately revised that the same typographical errors are found in each; for instance, the errata at the end have been allowed to remain without correction, and pp. 127 and 134 are reprinted by mistake 227 and 334. The book has become very valuable, because so few copies of it are known to exist. In one of Hearne's memoranda dated 1723, we read, "Horrox's posthumous works were printed by Wallis: they are now scarce. Mr. Whiteside, of the Museum, bought them several years agoe —but gave 7*s.* 6*d.* for them."

It has often been said to be a reflection upon our country that the writings of Horrox should have lain dormant for so long a time. As we have seen, it was upwards of twenty years before they were brought to light; and his beautiful dissertation upon the transit of Venus made its appearance in a foreign land. This was no doubt owing in part to the troubled state of the times. Political excitement and civil discord are not favorable to the advancement of literature and science. Moreover it should be remembered that Horrox was then unknown to fame. He lived in a remote part of the country, and died young. As soon as the value of his papers had been ascertained, measures were immediately adopted for their security, and eventually for their publication. It would indeed have been better if the account of the transit had been bound up with the rest of his posthumous works, according to the expressed wish of Flamsteed; but it seems that Wallis was under the impression at the time that a distinct edition of the *Venus* was about to be prepared by the Astronomer Royal, who was believed to be in possession of an autograph manuscript. No doubt he would willingly have included it in the volume if some good reason had not prevented him, for no one could have shewn greater zeal for the honor of Horrox, nor could have more deeply regretted that his celebrated observations should have been so long buried in obscurity. He says, "I cannot help being displeased that this valuable observation, purchaseable by no money, elegantly described, and prepared for the press, should have lain hid for two-and-twenty years, and that no one should have been found to take charge of so fair an offspring at its father's death, to bring to light a treatise of such importance to astronomy, and to preserve a work for our country's credit and for the advantage of mankind." The complaint is not entirely without foundation; but it is at all events a comfort to reflect that as soon as the manuscripts were discovered, Horrox's fame was endorsed by a society consisting of the most learned of his countrymen, that his writings were printed at public expense, and that his dissertation upon the transit received a graceful recognition from the leading astronomers of the continent of Europe.

Besides the manuscripts already mentioned, there are others of considerable interest, said to have formerly belonged to Flamsteed, which are now lodged in the library of the Greenwich Observatory. Of these we may enumerate,—Firstly, a transcript of the first twelve chapters of the *Venus in sole visa*, being a small book six inches high and four-and-a-half wide, containing fifty-eight clearly written pages, the last of which is not full, breaks off in the middle of the line, and is followed by several blank leaves. The account of the observations very nearly corresponds with that given by Hevelius. There are however no side-notes, a fact which confirms the belief that those attached to the printed edition formed no part of the original text. Although this document bears no date, the time when it was written may be concluded from a curious circumstance which we must not omit to notice. In the poem on the Telescope, inserted in the middle of the second chapter, there are some verses not to be found in the publication of Hevelius:

> "Et duplici nimium cœlesti a fonte remoti
> Tristia Saturni solatur lumina flamma."

Now Huygens first discovered a satellite of Saturn in the year 1665; Cassini discovered a second in October 1671, and a third in December 1672. These lines are therefore evidently an interpolation, since it was not known that Saturn had any satellite until twenty-four years after Horrox's death. They also prove that the manuscript is not of much authority, as it could not have been written for more than thirty years after the same event; and that although it belonged to Flamsteed, it is not the autograph which he was believed to possess, and from which it was thought he intended publishing a new and revised edition of the Venus. It may however be used with advantage for suggesting improved readings, and for making corrections in punctuation.—Secondly, a manuscript upon half sheets of old foolscap, ruled, and doubled so as to make a quarto eight inches high and six wide, which consists of three distinct parts, each paged separately, and headed as follows: (1) "*Jeremiæ Horroxii Præludium Astronomicum*," agreeing in substance with the tract of a similar name already stated to have been incorporated by Wallis, in the *Opera Posthuma*. Only the first book, "*De Motu solis*," has been commenced, having two chapters, namely, one entitled "*De parallaxi solis horizontali*," and another "*De refractione solis et syderum*." (2) "*Astronomiæ Lansbergianæ censura*," a short treatise, ending abruptly, the last line of which is written as if it had been intended to be continued on the following page. (3) "*Jeremiæ Horroxii Astronomiæ Lansbergianæ censura et cum Kepleriana Comparatio,*" which contains the *Prolegomena*, and other pieces found in the printed works. Upon comparing these two manuscripts with the *Opera Posthuma*, the general impression is, that they are in many places less full than the published text. Nevertheless they are extremely useful in throwing light on obscure passages, and in enabling us to form some idea of the manner in which Wallis arranged his materials. They are both in the same handwriting, which is certainly better than Flamsteed's, and totally different from his in character. They were probably penned by a regular transcriber; and it may be concluded, therefore, that they are of the same date.—Thirdly, an English manuscript called *Philosophical Exercises*, being a small book, about the size of the *Venus in sole visa*, divided into two parts, namely: (1) A discussion respecting the elliptical motions of the planets, and (2) Some more explicit rules upon the same subject. The sun's parallax is treated of nearly in the same way as that great question is discussed in the papers printed by Wallis. Towards the end there is "*A New Theory of the Moon*," which seems, from a comparison of the numbers employed, to have been the same as that adopted by Flamsteed; but this is only a conjecture, as the latter part of the document is very incomplete. This manuscript is evidently older than either of the other two, nor is there anything against the supposition of its having been written in the lifetime of Horrox. It is invested with peculiar interest, as being the only English composition of his in existence; and it is in general style more like an autograph than a transcription.

No monument was erected to the memory of Horrox, nor any mark set over his grave, for nearly two centuries after his death. In the year 1826, Mr. Holden, of Preston, delivered a course of lectures upon astronomy in Liverpool, and devoted the proceeds of one of his evenings to the erection of a suitable tablet, which was placed in St. Michael's Church, Toxteth Park. This was a proof of his appreciation of the merits of Horrox, and of his love for science; and it was an act which deserves general admiration. The monument is a handsome scroll of white marble, mounted on a black slab, having the appropriate representation of Venus crossing the Sun's disc, beneath which is the following inscription:

Venus in sole visa. Nov. 24, 1639.

IN MEMORY OF
JEREMIAH HORROX, ONE OF THE GREATEST
ASTRONOMERS THIS KINGDOM EVER PRODUCED;
BORN IN TOXTETH PARK IN 1619;
DIED IN 1641, AGED 22.

HIS OBSERVATIONS WERE MADE AT HOOLE,
EIGHT MILES FROM PRESTON, WHERE HE
PREDICTED, AND WAS THE FIRST PERSON
WHO SAW, THE TRANSIT OF VENUS
OVER THE SUN.

THIS MONUMENT WAS ERECTED BY
M. HOLDEN, ASTRONOMER
1826.

But the name of Horrox is not commemorated in his native place only; it is no less so in the parish of which he was a minister. The traditionary remembrance of the young astronomer which still exists at Hoole, began last year to assume a more substantial form. The Rev. Mr. Brickel, the present incumbent, naturally takes an interest in his fame, and as his successor in office, felt privileged to take measures for handing it down to posterity. Occupying the same pulpit Sunday after Sunday, he longed to identify Horrox with the parish to the end of time, by raising a lasting tribute to his memory. Hitherto there had been no record of his connection with Hoole; excepting that upon the old Church clock and sun-dial Horrox had inscribed the appropriate words "*ut hora, sic vita*," and "*sine sole sileo*," calculated to remind us of the shortness of life, and of our helplessness until the "Sun of Righteousness arise" upon the soul "with healing in His wings." With this view Mr. Brickel addressed to the gentlemen of influence in his neighbourhood, and to various scientific men throughout the country, a statement of the facts of the case, and asked their sympathy and assistance. The learned gave their testimony in favor of so distinguished a member of their brotherhood, and men of high position announced their readiness to contribute in furtherance of so laudable an undertaking. When sufficient funds were obtained, it was decided that the Church should be beautified, and enlarged by the erection of a chapel to be dedicated to the memory of Horrox which should contain thirty sittings free to the poor for ever. It was also agreed that a memorial window should be placed in it, together with a mural tablet having the following inscription:

JEREMIAH HORROCKS
BORN AT LIVERPOOL, EDUCATED AT CAMBRIDGE, THE CURATE
OF HOOLE,
DIED IN THE 22ND YEAR OF HIS AGE, 1641.

THE WISDOM OF GOD IN CREATION WAS HIS STUDY FROM EARLY YOUTH:
FOR HIS WONDERFUL GENIUS AND SCIENTIFIC KNOWLEDGE
MEN SPEAK OF HIM AS
"ONE OF ENGLAND'S MOST GIFTED SONS,"
"THE PRIDE AND BOAST OF BRITISH ASTRONOMY."
AMONGST HIS DISCOVERIES ARE—THE NEAREST APPROXIMATION TO THE
SUN'S PARALLAX.
THE CORRECT THEORY OF THE MOON, AND THE TRANSIT OF VENUS.

BUT THE LOVE OF GOD IN REDEMPTION WAS TO HIM A YET NOBLER THEME;
THE PREACHING OF CHRIST CRUCIFIED A YET HIGHER DUTY;
LOVING SCIENCE MUCH, HE LOVED RELIGION MORE;
AND TURNING FROM THE WONDERS OF CREATION TO THE GLORIES
OF THE CROSS, HE EXPRESSED THE RULE OF HIS LIFE
IN THESE MEMORABLE WORDS—
"*Ad majora avocatus, quæ ob hæc parerga neglig non decuit.*"

IN MEMORY OF ONE
SO YOUNG AND YET SO LEARNED,
SO LEARNED AND YET SO PIOUS,
THIS CHURCH IN WHICH HE OFFICIATED,
HAS BEEN ENLARGED AND BEAUTIFIED.

This was accordingly done; and the parish authorities have replaced the old dial and time-piece by a handsome clock, which is both an ornament to the church, and a convenience to the people. In this way the desire to do honor to Horrox has been crowned with success; and we can only trust that the blessing of God may rest upon the increased numbers who are now enabled to worship in His sanctuary.

In estimating the attainments of this remarkable young man, it must be remembered that we possess only a small portion of his writings, the bulk of them having unfortunately perished. His published works are but a part of what he wrote, and many of the tracts of which they are composed were left in an unfinished state. Hence some doctrines are treated systematically, whilst others are introduced here and there as occasion required. Omitting what might be inferred from a general survey of his papers, it will be sufficient for our purpose to mention such subjects only as are discussed in regular order. We must remember also that since he lived, more than two centuries have passed away, during which period a number of men have arisen, by whose genius and industry astronomy has been considerably developed. Our object is not to shew that he was abreast with the learning of the nineteenth century, but that he was greatly in advance of his own times; and that his exertions have in some measure contributed to elevate the science to its present proud position. The simple question to be answered is: What has been the practical value of his labors? What advantage were they to those who came after him? In other words: What has Horrox done for the improvement of astronomy?

The nature of his controversial papers has already been explained. Their object was to expose the vicious theories then prevailing, and to disseminate rational and correct views respecting the system of the universe. That his treatises against Lansberg and Hortensius were well calculated to effect this, there can be no doubt; but unfortunately they remained so long in an unpublished state that their usefulness was much impaired. Twenty years is a period of great importance in an era of progress. Nevertheless these papers were not unserviceable; as soon as they were printed, they were read with great interest, and passed through more than one edition. His observation of the transit of Venus was most valuable. No other person witnessed, with anything like success, the transit of 1639. By it he was enabled to correct the planet's elements and to prove, contrary to the received opinion, that her disc does not subtend an angle greater than one minute. He also estimated the sun's horizontal parallax more accurately than any one who came before him: it had previously been supposed to be at least two minutes, and even Kepler had stated it at 57″; but Horrox proved that it could not exceed 14″, which was within 1½″ of the value assigned to it by Halley sixty years afterwards. Horrox's reduction of the sun's parallax is very remarkable; for though he had not diminished it enough, Newton in the first edition of the *Principia* (1687) hesitated in following him so far. He said "I am not quite certain about the diameter of the earth as seen from the sun. I have assumed it to be 40″, because the observations of Kepler, Riccioli, and Vandelini do not allow of its being much greater. The observations of Horrox and Flamsteed make it somewhat less." He afterwards speaks of the apparent diameter of the earth as "about 24″, and therefore the parallax of the sun would be about 12″, very nearly as Horrox and Flamsteed had determined. But the diameter would agree better with the rule of this corollary if it were a little larger"—"*quasi* 24″, *adeoque parallaxis solaris quasi* 12″, *ut Horroccius et Flamstedius propemodum statuere. Sed diameter paulo major melius congruat cum regula hujus corollorarii.*" In the second edition of the *Principia* (1713) all this is omitted, and in a preceding corollary we read "the parallax of the sun from the most recent observations is about 10″." In the third edition he estimated it at 10½″. When it is remembered what expensive expeditions have been sent out from our country for the purpose of observing these transits, it is thought that the importance of the observation, and the conclusions derived from it will not seem to be over-rated. But as we have intimated, his fame chiefly rests upon the improvements he made in the lunar theory. His views upon this subject have been received with gratitude by the ablest astronomers. Newton's acknowledgment that he was the first to discover the motion of the moon to be in an ellipse about the earth, with the centre in the lower focus, has been already referred to; the exact words in the *Principia* are "*Horroccius noster lunam in ellipsi circum terram, in ejus umbilico inferiore constitutum, revolvi primus statuit:*" and upon comparing the different editions of the book, it will be seen that this statement was added to the second, and retained in the third. In his separate work, "*De mundi systemate,*" he speaks of Horrox's correction of the lunar theory in terms of great admiration: "There are many inequalities in the moon's motion not yet noticed by astronomers. They are all deducible from our principles, and are known to have a real existence in the heavens. This may be seen in the hypothesis of Horrox which is the most ingenious, and if I do not deceive myself, the most accurate of all:—*in Horroccii Hypothesi illâ ingeniosissimâ et ni fallor omnium accuratissima videre licet.*" Flamsteed declared his hypothesis for settling the movements of the moon to be the most exact that had ever been originated; and he did not even think it necessary to re-calculate the tables which Horrox, for want of time, had not verified to his own satisfaction. Halley, after speaking of the theories of various eminent men, says: "but that one alone of our Horrox which attributes to the moon's orbit a libratory motion of the apsides, and a variable eccentricity, seems to approach the truth of nature; for it represents the diameters more agreeably to observation, and shews her motion more accurately than any hypothesis which I have hitherto seen." We may further mention that Sir Isaac Newton largely availed himself of Horrox's suggestions to explain the general principles of perturbation, as laid down in the 66th proposition of the first book of the *Principia*. These improvements are so substantial that there is no difficulty in ascertaining the author to whom they are to be assigned. They stand out as a landmark in the history of the science. Taken in connection with his comments upon the subject of planetary motion, they prove that Horrox holds a prominent position amongst those who have succeeded in developing that great principle by which creation is held together. Few men are permitted to originate, to confirm, and to promulgate a great discovery. This is usually the work of successive generations. Each master-spirit pushes the enterprise a step further; and hence it is often difficult to decide who is fairly entitled to the credit. The final elucidation may be the result of an accumulated experience. The ground is first broken up, then the seed is sown, the tender plant is trained, and it grows and thrives, until some one more fortunate than the rest gathers the fruit. So it was with the principle of gravitation, the discovery of which cannot be wholly attributed to one man. It was, no doubt, reserved for the transcendent genius of Newton fully to define and to apply it; but the existence of such a power was known to others who came before him; and their ideas respecting it formed part of the data from which he drew his sublime conclusions. Thus Kepler had a considerable knowledge of the subject, and many of his conjectures have been substantiated. Dr. Gilbert published similar doctrines in this country, and gave them a more extended application. But Horrox, by his explanation of the perturbative influence of the sun, and by his illustration of celestial and projectile motion, unfolded the theory more completely than any of his predecessors. He seems to have perfectly understood the identity and universality of this unseen power; for he often tells us that the planets in their orbits are affected by it in the same manner as bodies upon the surface of the earth. His accurate views were at length adopted by Newton, and made the foundation of his philosophy. In proof of this compare the following passages:

"Just as by the force of gravity a projectile might describe an orbit, and revolve round the whole earth; so the moon, either by the force of gravity if it is endued with gravity, or by any other force urging it towards the earth, may be continually drawn thereto from a rectilineal path, and turned into her present orbit; and without such a force she cannot be retained in her orbit. If the force were less than it is, it would not cause her to deviate from a rectilineal course sufficiently: if it were greater, it would cause her to deviate too much, and draw her from her orbit towards the earth. It is therefore required to be of an exact amount; and it is the business of mathematicians to find the force which can accurately retain a body with a given velocity in any given orbit; and in like manner to find the curvilineal path into which a body going forth with a given velocity from any given place is turned from its rectilineal way by a given force."—*Newton Princip. Mathem. Def. V.*

"It is surely conceded by all that the motion of the planetary bodies is neither perfectly circular, nor perfectly uniform; for observations shew, beyond dispute, that the figure of the planetary orbits is elliptical or oval, and different from a circle: and the motion of a body in this ellipse is irregular being increased or diminished according to its distance from the sun. Physical causes are not wanting to shew that this movement is described by a sort of geometrical necessity. We may satisfy ourselves of the truth of this by an appeal to nature; for as a planet is moved by a magnetic impulse, why may not the same principle be exercised in other ways? A weight is thrown into the air: at first it rises quickly, then moves slowly, until at length it is stationary, and falls back to the earth with a velocity which continually increases. It thus describes a libratory movement. This movement arises from the impetus in a right line which has been imparted to it by your hand, together with the magnetic influence of the earth, which attracts all heavy things to itself, as a loadstone does iron. There is no need to dream of circles in the air, and I know not what, when we have the natural cause before our eyes; and as regards the motion of the planets which are subject to similar influences, what reason, I ask, is there to barter an explanation, the truth of which is comfirmed by so many examples in nature, for a fictitious dream of circles?" —*Jer. Hor. Op. Posth. Disp. VI. Cap. I.*

These paragraphs contain the same ideas expressed in different language. They both treat of the physical cause of curvilineal motion, which is explained to be the joint action of projectile and attractive forces; and they both speak of it as pervading the planetary system, and illustrate it by movements upon the surface of the earth. Now as Sir Isaac Newton is known to have been well acquainted with all that passed through Wallis' hands, he must have seen Horrox's treatise "*De Motu Syderum*," from which the above extract is taken; and he tells us himself that he had read his theory of the moon, in which the same principles are laid down. Without wishing to detract from the merits of one who, as an astronomer, has gained an immortal reputation, it is only right that it should be known that some of the leading doctrines upon which the philosophy of the *Principia* is built were first propounded by Horrox. Dr. Tatham in his "Chart and Scale of Truth," delivers his opinion upon this question in these words:

"That every philosopher has an absolute right to avail himself of the labors and discoveries of his predecessors, as a legacy freely given him, is a privilege which philosophy itself always claims. It is however a tribute justly due to the memory of this extraordinary genius, Mr. Horrox, whilst we regret the loss of many of his valuable works, to acknowlege from what has been saved, that he was principally instrumental in calling philosophy out of the regions of fictitious invention, and putting her on the investigation of the physical causes of things from experiments and observations; that he not only made the applications of projectile motion to the analogical illustration of celestial, but also assigned the forces of projective and attractive, on which all geometrical calculations are founded; and that, without injuring the immortal fame of his great successor, he may be fairly considered the forerunner of Newton."

We may conclude these observations upon the practical value of Horrox's labors by briefly remarking that he was the first to predict and observe the transit of Venus in 1639; to reduce the Sun's parallax nearly to what it has since been determined; to discover the orbit of the Moon to be an ellipse about the earth with the centre in the lower focus; to explain the causes of orbital motion; to ascertain the value of the annual equation with any degree of accuracy; to devise the beautiful experiment of the circular pendulum for illustrating the action of a central force; and to commence a regular series of tidal observations for the purpose of philosophical enquiry: besides all which, he effected improvements in different astronomical tables, recommended the adoption of decimal notation, detected the inequality in the mean motion of Jupiter and Saturn, and wrote his opinions upon the nature and movements of comets. That so much should have been achieved by so young a man, notwithstanding many disadvantages, may seem almost incredible; but if there is one fact connected with Horrox which, more than another, rests upon incontrovertible evidence, it is the age at which he died. This shews the lustre of his genius, and imparts a melancholy interest to his history. Those who have arrived at distinction in intellectual pursuits have generally done so early in life. Newton laid the foundation of his greatest discoveries before he had attained his thirtieth year; Byron expired at thirty-six; Pascal at thirty-nine; Mozart at thirty-five; and Raphael at thirty-seven; but Horrox's years were fewer still; they were not *twenty-two* in number. Such being the case, it is almost superfluous to say that he was gifted with the highest mental qualifications. As an instance of his extraordinary sagacity we may mention his early appreciation of Kepler's works which the philosophers who were contemporary with Horrox could not understand. Riccioli, Bouillaud, and others studied them to no purpose, whereas he embraced them at once. He speaks of Kepler as the "Prince of astronomers to whose discoveries alone all who understand the science will allow that we owe more than to those of any other person:" he says that he venerates his "sublime and enviably happy genius, and if necessary would defend to the utmost the Uranian citadel of the noble hero who has so far surpassed his fellows;" and he adds, "no one while I live shall insult his ashes with impunity." At the same time he took nothing upon trust, but carefully examined every theory that was propounded. Thus he writes, "The calculations of Lansberg and Longomontanus are false. Their principles and numbers are false. Kepler's hypotheses are true, and he seldom fails in his numbers." He possessed a habit of self-reliance; and we often find him complaining of the servility with which the astronomers of his day followed in each other's track without verifying by observation the doctrines that were handed down. In his speculations upon physical causes he was never at a loss for a new line of thought; but if it did not lead to a sound conclusion, it was dismissed as readily as it had been called forth. His power of reasoning out natural laws from the simple facts of common experience deserves especial notice. This is one of the greatest proofs of a philosophic mind. It is in fact to see more than is apparent to the common gaze. It enabled Newton to detect a great principle in the fall of an apple; and Galileo, whilst watching the swinging of a lantern in the Cathedral Church of Pisa, to conjecture that the oscillation of the pendulum might be turned to important purposes as a measure of time. Horrox beautifully expresses his belief in the harmony of nature; "Astronomy is natural and true. The sea is agitated with the winds; but the æther is clear and open, without wind or any other resistance. The bodies of the planets are solid and firm. Now as a slinger aims accurately, and projects his weapon with certainty, notwithstanding the resistance of the air, why may not the heavenly bodies, in like manner, rotate by an eternal law?" In short, Horrox possessed the spirit of a true philosopher; he was accustomed to generalize facts, to weigh probabilities, and to take the most ultimate views; and he improved to the utmost his noble powers by his unwearied industry and application. But scientific men are the most capable of forming an opinion of his merits, and to them we will appeal: Newton, and Foster of Gresham College, speak of him as "a genius of the very first rank;" and Sir Isaac, anticipating the publication of his works, expresses himself as "glad that the world will enjoy the writings of that excellent astronomer Horrox." Ferguson alludes to him as "our illustrious countryman;" Brinkley says that, had his life been spared, "his fame would probably have surpassed that of all his predecessors;" Herschel calls him "the pride and boast of British astronomy;" Dr. Whewell, the learned master of Trinity College, writes that, "he has attempted to do him justice;" Lord Brougham thinks that "nothing can be more clear than the great merit of Horrox, and the severe loss sustained by science from his early death;" Professor De Morgan says that "no monument is needed for the name of Horrox, for wherever Newton's *Principia* is known, there is his name known also;" and Professor Airy, the present Astronomer Royal, "joins warmly" in admiration of him. We will only add one more tribute to his praise: Grant, in his learned treatise upon physical astronomy, says that "Horrocks has exhibited in his researches such sagacity of thought and fertility of invention, such enlightened and judicious views on the various subjects which engaged his attention, and such unwavering confidence in the resources of his own mind," that, if he had remained on earth a few years longer, "his name would have been a household word for future generations."

Horrox was a poet as well as a philosopher. The verses which he has introduced in his account of the transit are very creditable, and evince a bold imagination combined with a judicious taste. They do not aim at being elaborate; indeed, he is so careless of detail, that by some his lines would be considered unpolished. Had he been a painter, his genius would have been impatient of the restraint which is implied by the speciality of arrangement found in the compositions of the pre-Raphaelite school; his ideas are strong and clear, and roughly delineated, whilst his metaphor somtimes borders upon exaggeration. But the sentiment which pervades his verse is delicate and refined. Enamoured of the heavens, he occasionally chooses poetry because it is the best vehicle for his passion; but in his advances he never forgets what is due to the society of the Muses. The Pierian spring gushes forth with unusual force, but its waters are always sweet and pure. His performances are powerfully conceived, freely executed, and are always in accordance with good taste. It is not often that poetic fancy and mathematical precision are so strongly developed in the same mind.

But intellect is of no value unless sanctified by grace. A man may be accounted a philosopher, he may explain the laws of Nature more successfully than any of his predecessors; but, if in his investigations of natural phenomena, he sees nothing but matter and motion, if he does not recognize the power, the wisdom, and the love of Him who creates and upholds, if he admires the work without admiring the workman, he is a philosopher "falsely so called." We are happy, therefore, before concluding this Memoir, to be able to bear testimony to Horrox's religious character. It is true that he left no theological papers; but this is not to be wondered at, as he was only permitted to exercise his ministration for so short a time. But if he did not write in the capacity of a clergyman, he thought and believed as a Christian; for we find sentiments introduced even in his most abstruse works, which show how much he lived under the influence of religion. A few passages in proof of this, besides those which have been already quoted, may be adduced. When he was about to enter upon the arduous task of correcting the Rudolphine tables, he says: "And may He who is the great and good God of astronomy, and the conservator of all useful arts, bless my unworthy efforts for His mercy's sake, and cause them to redound to the eternal glory of His name, and the advantage of mankind." In another place he writes that he will not despair of further discovery, "for I have been blessed by God's grace with such success, that even now I have something to be proud of." In his account of the transit of Venus where he speaks of being summoned, by his religious duties, from the observation which he knew he should never again have the opportunity of making, he draws a contrast between the importance of things temporal and things eternal which seems to express the general rule of his life and conversation, telling us that he was "called away to higher duties, which must not be neglected for these non-essentials." Would that this sentiment were more deeply felt by all who are engaged in the business of life! These isolated passages shew the spirit in which he did his work; but one of greater length has been preserved, where he speaks expressly of his own religious opinions and convictions. It of course partakes of the fanciful style of the schoolmen, and there is something in a typical representation of the Deity from which our more chastened thoughts necessarily shrink; but this fault belongs to the fondness for conceit which then prevailed, and must not blind us to the piety and humility of the writer. In connection with some crude philosophical speculations, we read: "I conclude that the eccentricity of the planets is caused by the contention between the suns magneticall (and always attractive) virtue, and the planets dulnes naturally desiring to rest unmoved, which dulnes, while the suns circular motion carrys the planet from the aphelium, is conquered, and so the planets motion increaseth in fastnes; but when the suns circular revolution doth recarry it backe toward the aphelium, the naturall torpor and dulnes increaseth, by the presence and nearnes of that place where it would rest.

"A right type may this be of mans dulnes to good, which is the more by how much a man more rests in himselfe, and is then onely quickned, when the Spirit of God (like the rays of the sun) doth draw our hearts, desirous to rest in themselves, and force them unwilling to follow Christ (as the planets follow the suns circumvolution, which begets a circular circumference), which following is the onely cause of our comming neer to god (as the suns circumference brings the planets towards itselfe). All which agrees excellently with that mysticall adumbration of the thrise sacred trinity in (those poor types of God as one calls them) round circles; wher the father (the center) doth beget the son (the circumference) by efflux of the spirite (the rays). Keplers astronomy differs from mine, as his religion: He gives the planets a divers nature (good and bad) that they may eyther come to the sun or fly away at their pleasure, or at least (as his second thoughts are) so dispose themselves (in spite of all the suns magneticall power) that the sun is bound to attract or expell them, according to that position, which themselves defend against all the suns labouring to incline the fibres. I, on the contrary, make the planet naturally to be averse from the sun, and desirous to rest in its owne place, caused by a materiall dulnes naturally opposite to motion, and averse from the sun, without eyther power or will to move to the sun of itselfe. But then the sun by its rays attracts, and by its circumferentiall revolution carrys about the unwilling planet, conquering that naturall selfe rest that is in it, yet not so far but that the planet doth much abate and weaken this force of the sun, as is largely disputed afore. So just do the papists, whose free will to good or bad, can by its owne strength, go to God or fly from him, or at least so frame their own actions, as that God is bound to save them or damn them volens nolens. But I will confesse myselfe not equally composed of good and bad, that myselfe may give eyther flesh or spirit the upper hand, but rather wholly desirous to rest in my selfe, wholly averse from God, and therefore justly deserve (as the fixed stars from the sun) to be blown away from God in infinitum, but that God by his Sons taking on him mans nature, and the undeserved inspirition of his spirit, doth quicken this dulnes, nay deadnes of my nature, yet still, ah me! how doth it choke and weaken those operations! If any one thinke all this but an idle conceit, I must tell him he doth too rashly deride that booke of creatures, that voyce of the heavens which is heard in all the world, and wherein without question God hath instamped more mysterys than the lazy witts of men, more ready to slight than amend any speculation, are ordinarily aware of. Shall we thinke that he who was content to shadow out these mysterys with the poor blood of buls and goats, will disdain to have them typified in the more glorious bodys of the stars and motions of the heavens; which David accounted such cleare Emblems of Gods glory that he goes from speaking of the light of the sun, unto Gods law, as if the subiect were still the same, without any conclusion to the first, or introduction to the latter. For my part I must ever thinke that God created all other things, as well as man, in his own image, and that the nature of all things is one, as God is one, and therefore an harmonicall agreeing of the causes of all things, if demonstrated, were the quintessence of most truly naturall philosophy.

Sic itur ad astra,
Repet hum: quicunque velit."

The curious illustrations in this extract will easily be pardoned, when it is remembered that they were in accordance with the phraseology of the day. In later times, Wallis imagined that the doctrine of the Trinity might be exemplified by the three dimensions of a cube; and even the theological treatises of the first half of the seventeenth century abound with expletives which would now be considered unsuitable to the solemnity of the subject. The passage breathes sentiments of the purest piety, and it is gratifying to know that Horrox had such clear views of evangelical truth. The cause of religion is strengthened when men of intellect range themselves on the Lord's side; and the sneer of the scoffer is repressed, whose specious arguments might otherwise unsettle the faith of the weaker brethren, and throw poison into the waters of life. How often do people take exception at some statement of scripture because it appears to them to be irreconcileable with the fresh dis coveries of science; and although the point in dispute may be comparatively unimportant, they magnify its proportions, until the great principles of the Bible are completely put out of sight: whereas, by deferring their judgment for awhile, it would be seen that such discoveries, if true in themselves, are not opposed to the teaching of Revelation. For it should be remembered that all truth proceeds from one great source: it has its foundation in the character of God. Science and religion therefore can never be hostile to each other; because they both work up to a common centre. The beneficence and order which are so conspicuous in the constitution of the universe were made known in the pages of scripture, generations before the physical sciences were cultivated. They are particularly conspicuous in the plan of redemption. In this respect, the arrangements of Providence resemble those of Grace. At one time it was thought that the inequalities in the movements of the heavenly bodies would prove fatal to the establishment of the principle of gravitation; instead of which, upon further investigation, it was found, that so far from being a violation of the general law, they afforded a remarkable confirmation of it. In like manner we read in the Gospel, that God can be "just, and yet the justifier of him which believeth in Jesus." This doctrine would not have been deemed possible by the sages of old, and when first preached, it was a stumbling-block to many; professing themselves to be wise they became fools; but a patient and unprejudiced examination convinces us that it is not only agreeable to the perfections of God, but even throws a lustre on His character, to which mankind before were strangers. Religion and science then are only different departments of truth; they can have no conflicting interests. The subject of this Memoir was eminent in the pursuit of both. He saw the work of a Father's hand in the stars of heaven, the flowers of the field, the cattle upon the hill-side, the attributes of man, and in the rich provision that has been made for every endangered heir of glory. He knew that even the evil that is in the world is a part of the general plan of administration; that sin is permitted to exist only for the manifestation of a much more abounding grace; and that the present dispensation is introductory to one more perfect and more enduring, when the irregularities which now perplex us shall be seen to have been ordained in wisdom and love. Thus whilst he took pleasure in following up the path of discovery, and sought to carry the line and compass to the utmost boundaries of science, he was careful to study and to practise beyond everything the laws of God's spiritual kingdom. and thus to prepare for the future world of light and happiness. In a word, the greatest proof of his intelligence was, that he lived and acted for Eternity.

"While yet on earth the youthful pastor trod,
He read the word and traced the works of God;
The courses of the stars prophetic saw,
Unwound their order, and defined their law.
And yet a loftier view his eye could scan—
For this lost world salvation's glorious plan—
The firmament of souls redeemed from night,
The centre Jesus, and the circle light.
A Sage's love, a young Apostle's zeal,
The head to reason, and the heart to feel—
With truth and mercy graced the preacher's tongue,
And o'er his life a holy radiance flung.
That meteor—life, soon lost to vision here,
Now shines unclouded in a glorious sphere;
Yet here its light his bright example gives,
And here in fame undying Horrox lives."

The Transit of Venus over the Sun:

OR

AN ASTRONOMICAL TREATISE

ON

THE CELEBRATED CONJUNCTION

OF

VENUS AND THE SUN

On the 24th of November, 1639.

By JEREMIAH HORROX.

CHAPTER I.

The occasion, excellence, and utility of the Observation.

Soon after the commencement of my astronomical studies, and whilst preparing for practical observation, I computed the Ephemerides of several years, from the continuous tables of Lansberg. Having followed up the task with unceasing perseverance, and having arrived at the point of its completion, the very erroneous calculation of these tables, then detected, convinced me that an astronomer might be engaged upon a better work. Accordingly I broke off the useless computation, and resolved for the future with my own eyes to observe the positions of the stars in the heavens; but lest so many hours spent on Lansberg should be entirely thrown away, I made use of my Ephemerides in ascertaining the positions of the distant planets, so that I was enabled to predict their conjunctions, their appulses to the fixed stars, and many other extraordinary phenomena. Delighted for the time with such a foretaste of the science, I took great pains carefully to prepare myself for further observation.

Whilst thus engaged, I received my first intimation of this remarkable conjunction of Venus with the Sun; and I regard it as a very fortunate occurrence, inasmuch as about the beginning of October, 1639, it induced me, in expectation of so grand a spectacle, to observe with increasd attention. I pardon, in the meantime, the miserable arrogance of the Belgian astronomer, who has overloaded his useless tables with such unmerited praise, and cease to lament the misapplication of my own time, deeming it a sufficient reward that I was thereby led to consider and to foresee the appearance of Venus in the Sun. But on the other hand, may Lansberg forgive me that I hesitated to trust him in an observation of such importance; and, from having been so often deceived by his pretension to universal accuracy, that I disregarded the general reception of his tables. Besides, I thought it my duty to consult other calculations, especially those of Rudolphi, which Hortensius has vainly labored to depreciate. Daily experience indeed convinces me that what Lansberg says (whether with less modesty or truth I know not) of his own tables may be affirmed with propriety of Kepler's, namely, that they are superior to all others.

"Quantum lenta solent inter viburna cupressi."

The more accurate calculations of Rudolphi very much confirmed my expectations; and I rejoiced exceedingly in the hope of seeing Venus, the rarity of whose appearance in conjunction with the Sun had induced me to pay less attention to the more common phenomena of the same kind visible in the planet Mercury; for though hitherto these phenomena have been observed on one occasion only, the science of astronomy holds out to us the assurance that they will, even in our time, frequently appear.

But lest a vain exultation should deceive me, and to prevent the chance of disappointment, I not only determined diligently to watch the important spectacle myself, but exhorted others whom I knew to be fond of astronomy to follow my example; in order that the testimony of several persons, if it should so happen, might the more effectually promote the attainment of truth; and because by observing in different places, our purpose would be less likely to be defeated by the accidental interposition of the clouds or any fortuitous impediment.

The chance of a clouded atmosphere caused me much anxiety; for Jupiter and Mercury were in conjunction with the Sun almost at the same time as Venus. This remarkable assemblage of the planets, (as if they were desirous of beholding, in common with ourselves, the wonders of the heavens, and of adding to the splendour of the scene), seemed to forebode great severity of weather. Mercury, whose conjunction with the Sun is invariably attended with storm and tempest, was especially to be feared. In this apprehension I coincide with the opinion of the astrologers, because it is confirmed by experience; but in other respects I cannot help despising their more than puerile vanities.

I have thought it right, independently of the remarks upon the planets which I have elsewhere made, to publish a separate treatise upon this observation, on account of its great practical utility and excellence above all others, which I trust I may be permitted to set forth without being accused of ostentation.

In the first place, I found that it was well suited to correct the mean motion of Venus, on account of two advantages which other observations do not possess.

The one consists in the difficulty which might be occasioned by the parallax of the orbit, or the second equation, being removed from this observation. I speak in accordance with the opinion of

Copernicus, whom alone I shall follow in his general hypotheses. The conjunction placing the bodies of the sun, of the earth, and of the planet herself in one line has removed all possibility of deception from a spectacle which in other positions presents difficulties scarcely possible to overcome.

The other advantage results from the proximity of Venus to the earth, and her convenient situation as respects the sun, whence it happens that one minute in her longitude alters her apparent situation nearly three minutes. If therefore on the other hand, we can observe her apparent place within a minute, it is clear that we shall ascertain her real longitude in her orbit within the third part of a minute; whereas when the planet is in other situations, a whole degree scarcely affects the apparent place of her longitude, especially in her greatest distances from the sun, when observations of her are most frequently and correctly made; moreover both these and other observations plainly prove that the mean motion of Venus has never yet been determined by astronomers with sufficient accuracy.

In the second place, no other observation shews

so correctly the longitude of the node of Venus; for the telescope which I employed on this occasion is much more accurate than those generally used. Neither have I depended altogether upon the latitude of the fixed stars, with regard to which there might be some doubt, but have calculated from the sun itself, which is always necessarily fixed in the Ecliptic. Moreover there is an additional circumstance in the very great visible inclination of the orbit, by which, the apparent latitude being rapidly changed, the distance of Venus from the node is more minutely ascertained; one minute of observed latitude determining the longitude of the node to the tenth part of a degree; upon this point, however, it is right to add that modern astronomers are divided.

But especially would I call the attention of the reader to the surprising minuteness of Venus' apparent diameter; even though Gassendi has already bespoken the admiration of astronomers, by pointing out a similar peculiarity with respect to Mercury; and though I am not the first to notice this circumstance, I can at all events confirm it. By another and a very striking proof it will be seen how much we are liable, in estimating the diameters of the planets, to be deceived by their refraction.

Influenced by these reasons, and following the example of Gassendi, I have drawn up an account of this extraordinary sight, trusting that it will not prove less pleasing to astronomers to contemplate Venus than Mercury, though she be wrapt in the close embraces of the sun;

Vinclisque nova ratione paratis
Admisisse Deos.

Hail! then, ye eyes that penetrate the inmost recesses of the heavens, and gazing upon the bosom of the sun with your sight-asissting tube, have dared to point out the spots on that eternal luminary! And thou too, illustrous Gassendi, above all others, hail! thou who, first and only, didst depict Hermes' changeful orb in hidden congress with the sun. Well hast thou restored the fallen credit of our ancestors, and triumphed o'er the inconstant Wanderer. Behold thyself, thrice celebrated man! associated with me, if I may venture so to speak, in a like good fortune. Contemplate, I repeat, this most extraordinary phenomenon, never in our time to be seen again! the planet Venus drawn from her seclusion, modestly delineating on the sun, without disguise, her real magnitude, whilst her disc, at other times so lovely, is here obscured in melancholy gloom; in short, constrained to reveal to us those important truths, which Mercury, on a former occasion, confided to thee.

How admirably are their destinies appointed! How wisely have the decrees of Providence ordered the several purposes of their creation! Thou, a profound Divine, hast honored the patron of wisdom and learning; whilst I, whose youthful days are scarce complete, have chosen for my theme the Queen of love, veiled by the shade of Phœbus' light!

CHAPTER II.

Account of the Observation.

WHILST I was meditating in what manner I should commence my observation of the planet Venus so as effectually to realize my expectations, the recent and admirable invention of the telescope afforded me the greatest delight, on account of its singular excellence and superior accuracy above all other instruments. For although the method which Kepler recommends in his treatise on Optics, of observing the diameter and eclipses of the sun through a plain aperture without the aid of glasses, is very ingenious, and in his opinion, on account of its freedom from refraction, preferable to the telescope; yet I was unable to make use of it, even if I had wished to do so, inasmuch as it does not shew the sun's image exactly, nor with sufficient distinctness, unless the distance from the aperture be very great, which the smallness of my apartment would not allow. Moreover I was afraid to risk the chance of losing the observation; a misfortune which happened to Schickard, and Mögling, the astronomer to the Prince of Hesse, as Gassendi tells us in his *Mercury:* for they, expecting to find the diameter of Mercury greater than it was reasonable to anticipate, made use of so large an aperture that it was impossible to distinguish the planet at all, as Schickard himself has clearly proved; and even though Venus gave promise of a larger diameter, and thereby in some measure lessened this apprehension, and I was able to adapt the aperture to my own convenience, yet in an observation that could never be repeated, I preferred encountering groundless fears to the certainty of disappointment. Besides, I possessed a telescope of my own of such power as to shew even the smallest spots upon the sun, and to enable me to make the most accurate division of his disc; one which, in all my observations, I have found to represent objects with the greatest truth. This kind of instrument therefore I consider ought always to be preferred in such experiments. As soon as its usefulness became known to me, I eulogized it in the following lines:

Divine the hand which to Urania's power
Triumphant raised the trophy, which on man
Hath first bestowed the wondrous tube by art
Invented, and in noble daring taught
His mortal eyes to scan the furthest heavens.
Whether he seek the solar path to trace,
Or watch the nightly wanderings of the moon

Whilst at her fullest splendour, no such guide
From Jove was ever sent, no aid like this
In brightest light such mysteries to display;
Nor longer now shall man with straining eye
In vain attempt to seize the stars. Blest with this
Thou shalt draw down the moon from heaven, and give
Our earth to the celestial spheres, and fix
Each orb in its own ordered place to run
Its course sublime in strict analogy.
For whilst thou see'st the lunar disc display
Such rocks and ocean-depths unfathomable,
What powers prevent thy sight of worlds celestial
From tracing all their semblance to this earth?
This hand divine, right bold Copernicus,
Supplies fresh arms to vindicate thy cause,
Supporting thee who dared to make the worlds
Revolve by laws unchangeable, it clothes
The hosts of heaven with earthly forms, and bids
The earth itself to claim the second place
Below the sun, a rival to the stars
That hold their stations in the realms of space.
Forbidding more the senseless crowd to rule
O'er minds whose high-aspiring thoughts shall soon
Surpass the utmost bounds of ancient lore,
Its powers disperse the troop that know no rule
But texts too vainly taught by him who gave
Such lasting honors to Stagira's name;

They tear to shreds a thousand fancied laws
That truth deface like spots upon the sun,
And send the tomes that else might lead astray
A fitting present to the moths and worms.
This prying tube too shews fair Venus' form
Clad in the vestments of her borrowed light,
While the unworthy fraud her crescent horn
Betrays. Though bosomed in the solar beams
And by their blaze o'erpowered, it brings to view
Hermes and Venus from concealed retreats;
With daring gaze it penetrates the veil
Which shrouds the mighty ruler of the skies,
And searches all his secret laws. O! power
Alone that rivalest Promethean deeds!
Lo, the sure guide to truth's ingenuous sons!
Where'er the zeal of youth shall scan the heavens,
O may they cherish thee above the blind
Conceits of men, and the wild sea of error
Learning the marvels of this mighty Tube!

Having attentively examined Venus with my instrument, I described on a sheet of paper a circle whose diameter was nearly equal to six inches, the narrowness of the apartment not permitting me conveniently to use a larger size. This however admitted of a sufficiently accurate division; nor could the arc of a quadrant be apportioned more exactly, even with a radius of fifty feet, which is as great an one as any astronomer has divided; and it is in my opinion far more convenient than a larger, for although it represents the sun's image less, yet it depicts it more clearly and steadily. I divided the circumference of this circle into 360° in the usual manner, and its diameter into thirty equal parts, which gives about as many minutes as are equivalent to the sun's apparent diameter: each of these thirty parts was again divided into four equal portions, making in all one hundred and twenty; and these, if necessary, may be more minutely subdivided; the rest I left to ocular computation, which, in such small sections, is quite as certain as any mechanical division. Suppose then each of these thirty parts to be divided into 60″, according to the practice of astronomers. When the time of the observation approached, I retired to my apartment, and having closed the windows against the light, I directed my telescope, previously adjusted to a focus, through the aperture towards the sun and received his rays at right angles upon the paper

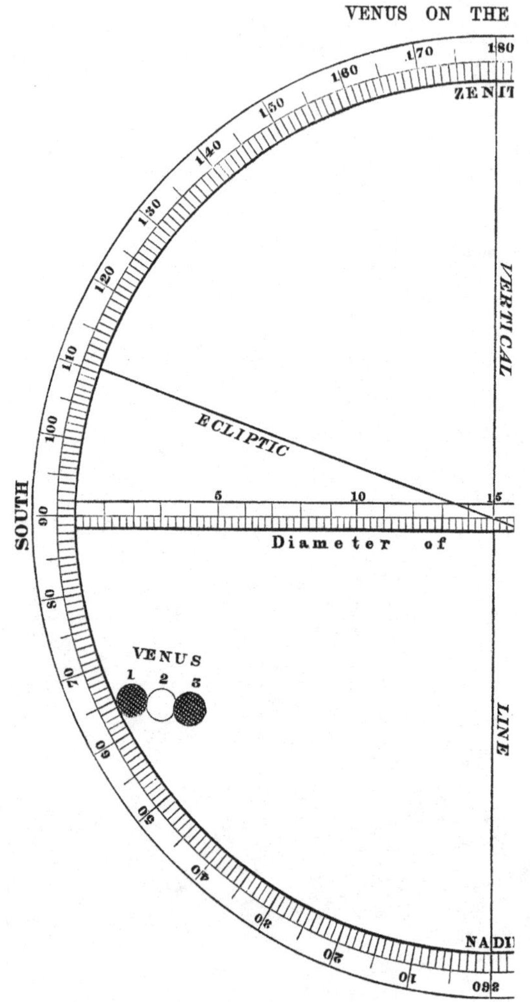
VENUS ON THE
ZENIT
VERTICAL
LINE
NADI
SOUTH
ECLIPTIC
Diameter of
VENUS
1 2 3
5 10 15

SUN'S DISC.

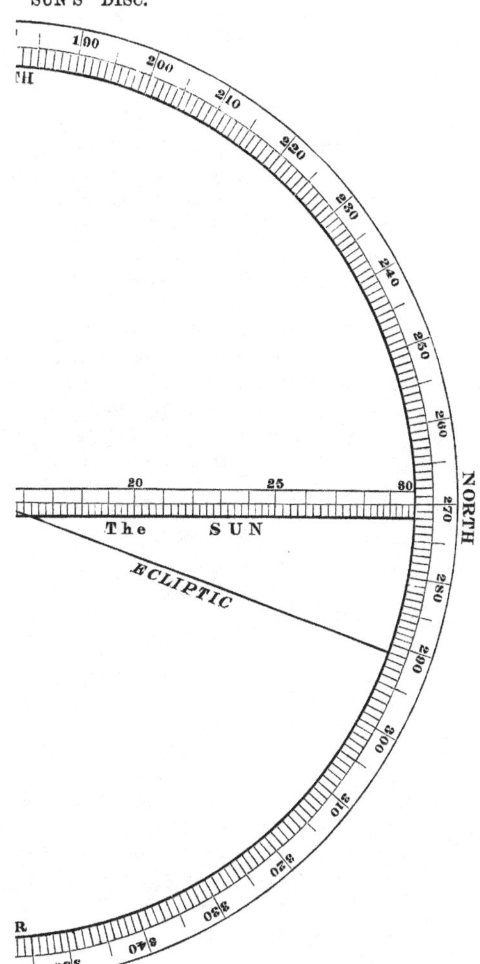

already mentioned. The sun's image exactly filled the circle, and I watched carefully and unceasingly for any dark body that might enter upon the disc of light.

Although the corrected computation of Venus' motions which I had before prepared, and on the accuracy of which I implicitly relied, forbad me to expect anything before three o'clock in the afternoon of the 24th ; yet since, according to the calculations of most astronomers, the conjunction should take place sooner, by some even on the 23rd, I was unwilling to depend entirely on my own opinion which was not sufficiently confirmed, lest by too much self-confidence I might endanger the observation. Anxiously intent therefore on the undertaking through the greater part of the 23rd, and the whole of the 24th, I omitted no available opportunity of observing her ingress. I watched carefully on the 24th from sunrise to nine o'clock, and from a little before ten until noon, and at one in the afternoon, being called away in the intervals by business of the highest importance which, for these ornamental pursuits, I could not with propriety neglect. But during all this time I saw nothing in the sun except a small and common spot, consisting as it were of three points at a distance from the centre towards the left, which I noticed on the preceding and following days. This evidently had nothing to do with Venus. About fifteen minutes past three in the afternoon, when I was again at liberty to continue my labors, the clouds, as if by divine interposition, were entirely dispersed, and I was once more invited to the grateful task of repeating my observations. I then beheld a most agreeable spectacle, the object of my sanguine wishes, a spot of unusual magnitude and of a perfectly circular shape, which had already fully entered upon the sun's disc on the left, so that the limbs of the Sun and Venus precisely coincided, forming an angle of contact. Not doubting that this was really the shadow of the planet, I immediately applied myself sedulously to observe it.

In the first place, with respect to the inclination, the line of the diameter of the circle being perpendicular to the horizon, although its plane was somewhat inclined on account of the Sun's altitude, I found that the shadow of Venus at the aforesaid hour, namely fifteen minutes past three, had entered the Sun's disc about 62° 30′, certainly between 60° and 65°, from the top towards the right. This was the appearance in the dark apartment; therefore out of doors beneath the open sky, according to the law of optics, the contrary would be the case, and Venus would be below the centre of the sun, distant 62° 30′ from the lower limb, or the nadir, as the Arabians term it. The inclination remained to all appearance the same until sunset, when the observation was concluded.

In the second place, the distance between the centres of Venus and the Sun I found, by three observations, to be as follows:—

	The Hour.		Distance of the Centres.
At	3 . 15	by the clock.	14′ 24″
„	3 . 35	„	13′ 30″
„	3 . 45	„	13′ 0″
„	3 . 50 the apparent sunset.		

The true setting being 3.45. and the apparent about 5 minutes later, the difference being caused by refraction. The clock therefore was sufficiently correct.

In the third place, I found after careful and repeated observation, that the diameter of Venus, as her shadow was depicted on the paper, was larger indeed than the thirtieth part of the solar diameter, though not more so than the sixth, or at the utmost the fifth, of such a part Therefore let the diameter of the Sun be to the diameter of Venus as 30′ to 1′ 12″. Certainly her diameter never equalled 1′ 30″, scarcely perhaps 1′ 20″, and this was evident as well when the planet was near the Sun's limb, as when far distant from it.

This observation was made in an obscure village where I have long been in the habit of observing, about fifteen miles to the north of Liverpool, the latitude of which I believe to be 53° 20′, although by the common maps it is stated at 54° 12′, therefore the latitude of the village will be 53° 35′, and the longitude of both 22° 30′ from the Fortunate Islands, now called the Canaries. This is 14° 15′ to the west of Uraniburg in Denmark, the longitude of which is stated by Brahé, a native of the place, to be 36° 45 from these Islands.

This is all I could observe respecting this celebrated conjunction, during the short time the Sun remained in the horizon: for although Venus continued on his disc for several hours, she was not visible to me longer than half-an-hour, on account of his so quickly setting. Nevertheless, all the observations which could possibly be made in so short a time, I was enabled, by Divine Providence, to complete so effectually that I could scarcely have wished for a more extended period. The inclination was the only point upon which I failed to attain the utmost precision; for, owing to the rapid motion of the Sun, it was difficult to observe with certainty to a single degree, and I frankly confess, that I neither did nor could ascertain it. But all the rest is sufficiently accurate, and as exact as I could desire.

CHAPTER III.

What others observed, or might have observed, of this Conjunction.

When first I began to attend to this Conjunction, I not only determined myself to watch diligently an appearance so important, but invited others also whom I knew to be interested in astronomy to do the same, in order that the testimony of many observers, should it so happen, might more firmly establish the truth ; and especially because, if observations were made in different places, our expectations would be less likely to be frustrated by a cloudy sky or any other obstacle. I wrote therefore immediately to my most esteemed friend William Crabtree, a person who has few superiors in mathematical learning, inviting him to be present at this Uranian banquet, if the weather permitted; and my letter, which arrived in good time, found him ready to oblige me; he therefore carefully prepared for the observation, in a manner similar to that which has been mentioned. But the sky was very unfavorable, being obscured during the greater part of the day with thick clouds; and as he was unable to obtain a view of the Sun, he despaired of making an observation, and resolved to take no further trouble in the matter. But a little before sunset, namely about thirty-five minutes past three, certainly between thirty and forty miuntes after three, the Sun bursting forth from behind the clouds, he at once began to observe, and was gratified by beholding the pleasing spectacle of Venus upon the Sun's disc. Rapt in contemplation, he stood for some time motionless, scarcely trusting his own senses, through excess of joy; for we astronomers have as it were a womanish disposition, and are overjoyed with trifles and such small matters as scarcely make an impression upon others; a susceptibility which those who will may deride with impunity, even in my own presence, and, if it gratify them, I too will join in the merriment. One thing I request: let no severe Cato be seriously offended with our follies; for, to speak poetically, what young man on earth would not, like ourselves, fondly admire Venus in conjunction with the Sun, "pulchritudinem divitiis conjunctam"? But to return, he from his ecstacy, and I from my digression. In a little while, the clouds again obscured the face of the Sun, so that he could observe nothing more than that Venus was certainly on the disc at the time. What he actually saw in so short a space was as follows:

In the apartment, Venus occupied the right side of the Sun, being higher than its centre, and therefore in the heavens lower and on the left. She was distant at the aforesaid hour, namely thirty-five minutes past three. a sufficiently appreciable space from the Sun's left limb; but Crabtree's opportunity was so limited that he was not able to observe very minutely either the distance itself, or the inclination of the planet. As well as he could guess by his eye, and to the best of his recollection, he drew upon paper the situation of Venus, which I found to differ little or nothing from my own observation; nor indeed did he err more than Apelles himself might have done in making so rapid a sketch. He found the diameter of Venus to be seven parts, that of the Sun being two hundred, which, according to my calculations, gives about 1′ 3″.

This observation was made near Manchester, called by Antoninus *Mancunium* or *Manucium*, the latitude of which Mr. Crabtree makes 53° 24′, and the common tables 45° 15′; the longitude 23° 15′, or three minutes of time to the east of Liverpool, from which it is distant twenty-four miles.

I wrote also of the expected transit to my

younger brother, who then resided at Liverpool, hoping that he would exert himself on the occasion. This indeed he did, but it was in vain; for on the 24th, the sky was overcast, and he was unable to see anything, although he watched very carefully. He examined the Sun again on the following day which was somewhat clearer; but with no better success, Venus having already completed her transit.

I hope to be excused for not informing other of my friends of the expected phenomenon, but most of them care little for trifles of this kind, preferring rather their hawks and hounds, to say no worse; and although England is not without votaries of astronomy, with some of whom I am acquainted, I was unable to convey to them the agreeable tidings, having myself had so little notice. If others, without being warned by me, have witnessed the transit, I shall not envy their good fortune, but rather rejoice, and congratulate them on their diligence. Nor will I withhold my praise from any who may hereafter confirm my observations by their own, or correct them by anything more exact. Let us then briefly consider what assistance may be expected from others.

In the space of half-an-hour, Venus advanced towards the centre of the Sun a distance of 1′ 24″; of course, therefore, in twenty-six minutes she had travelled to the extent of her own diameter, namely 1′ 12″; that is, as much as, at the first observation at fifteen minutes past three, the antecedent limb of Venus had passed over the Sun's limb; therefore forty-nine minutes past two was the commencement of her eclipse.

At Uraniburg, where there was formerly an observatory under Tycho, this would be forty-six minutes past three, but the Sun set there at half-past three, which is sixteen minutes before the commencement of the eclipse; therefore nothing could have been observed, even should astronomy not have perished with its patron, and some should be yet remaining who, having leisure for the pursuit, sustain the ancient credit of Uraniburg.

At Goesa, in Zealand, where Lansberg lately flourished, it commenced at fourteen minutes past three, and the Sun set at fifty-five minutes past three, consequently it might have been seen there. But no one excepting Lansberg and his friend Hortensius, both of whom I hear are dead, would trouble themselves about the matter; nor is it probable that, if living, they would be willing to acknowledge a phenomenon which would convict their much-vaunted tables of gross inaccuracy.

At Hesse Cassel the eclipse began at thirty-three minutes past three, the Sun set at fifty-five minutes past three. Providentially, Mr. Mögling would be prepared for the conjunction with his telescope, or at least with a tube furnished with a narrower aperture than that which was formerly used in observing Mercury; if indeed there is sufficient leisure in Germany to attend to subjects of so trivial a nature to the neglect of more important affairs.

At Paris, where Gassendi observed the conjunction of Mercury with the Sun, the transit was to be seen a little later than with us; for the first entry of Venus upon the Sun's disc took place at six minutes past three, whilst the true time of sunset was eight minutes past four, and the apparent at twelve minutes past four, therefore

Venus was visible in the Sun for more than an hour. Hence we shall consider Gassendi very fortunate if he have found her no less accessible than Mercury; and that neither unfavorable weather nor inadvertence, of which it would be wrong to accuse so celebrated an astronomer, deprived him of the opportunity.

In short, Venus was visible in the Sun throughout nearly the whole of Italy, France, and Spain; but in none of those countries during the entire continuance of the transit.

But America!

O fortunatos nimium bona si sua norint!

Venus! what riches dost thou squander on unworthy regions which attempt to repay such favors with gold, the paltry product of their mines. Let these barbarians keep their precious metals to themselves, the incentives to evil which we are content to do without. These rude people would indeed ask from us too much should they deprive us of those celestial riches, the use of which they are not able to comprehend. But let us cease this complaint, O Venus! and attend to thee ere thou dost depart.

"Why beauteous Queen desert thy votaries here?
Ah! why from Europe hide that face divine,
Most meet to be admired? on distant climes
Why scatter riches? or such splendid sights
Why waste on those who cannot prize their value?
Where dost thou madly hasten? Oh! return:
Such barbarous lands can never duly hail
The purer brightness of thy virgin light.
Or rather here remain: secure from harm,
Thy bed we'll strew with all the fairest flowers;
Refresh thy frame, by labors seldom tried,
Too much oppressed; and let that gentle form
Recline in safety on the friendly couch.
But ah! thou fliest! And torn from civil life,
The savage grasp of wild untutored man
Holds thee imprisoned in its rude embrace.
Thou fliest, and we shall never see thee more,
While heaven unpitying scarcely would permit
The rich enjoyment of thy parting smile.
Oh! then farewell thou beauteous Queen! thy sway
May soften nations yet untamed, whose breasts
Bereft of native fury then shall learn
The milder virtues. We with anxious mind
Follow thy latest footsteps here, and far
As thought can carry us; my labors now
Bedeck the monument for future times
Which thou at parting left us. Thy return
Posterity shall witness; years must roll
Away, but then at length the splendid sight
Again shall greet our distant children's eyes."

CHAPTER IV.

It is proved that the spot observed in the Sun's disc was really Venus.

THE most skilful astronomers in their observation of Mercury have been frequently deceived; firstly, those, who in the time of Charlemagne, on the 16th of April in the year 807, believed that the transit of Mercury over the Sun continued eight days: secondly, Averrhoes, who says in the *Ptolemaic Paraphrase*, that he recollected to have seen something of a darkish appearance, and subsequently found by the numbers that the conjunction of Mercury and the Sun had been predicted; he flourished about the year 1160 of the christian era: thirdly, Kepler himself, the most learned astronomer that ever lived, was greatly deceived on the 18th of May 1607. All these having seen spots on the Sun's disc, an appearance not understood in those days, rashly concluded them to be the planet Mercury; but they were evidently misled, as circumstances afterwards proved.

Are we then similarly deceived, and do we mistake an ordinary spot for Venus?

Verily since this may be doubted, as well by some who are unacquainted with the heavens except from books, as by others who are learned and practical astronomers; and lest our labor should be in vain, it may be worth while, before further prosecuting the enquiry, to prove in a satisfactory manner that the planet Venus was the actual cause of this appearance.

Firstly, perchance there may be some who believe that neither Venus nor Mercury could ever be seen in the Sun, although they might be upon his disc; such, for instance, as suppose that all the heavenly bodies shine with their own light, and are neither opaque nor cast a shadow like the Earth and Moon.

Secondly, others who, trusting to the astronomical tables which they imagine as accurate as their authors describe them to be, easily give way to the same opinion, and deny that any real transit took place on either the hour or the day we have specified; nor will they allow themselves to be persuaded that tables, boasting so confidently of their own accuracy, could possibly err to the extent of a whole day, or miscalculate the situation of Venus by several degrees.

But, thirdly, they will be the most astonished who, having contemplated this beautiful planet, which on a clear evening they think may even vie with the Moon, shall learn from us her surprising minuteness; and when they are told that the common opinion of astronomers makes the diameter of Venus equal to two-fifths of that of the Sun, that is, ten times greater than we have actually found it, they may possibly conclude that we have been deceived by an ordinary spot, and blinded by the desire of dignifying it with the name of Venus.

Let others fear such a conclusion: for myself, what I saw with my own eyes in the heavens, supplied me with sufficient evidence of the certainty of the observation, almost all the circumstances of which I had predicted to my friends; and I silently congratulate myself that my correction of the motion of Venus, which I had not before sufficiently appreciated, has been confirmed beyond my utmost hopes. In order to satisfy the doubts of others I make the following remarks :—

Firstly, there is no occasion for any one to be misled because Venus was deprived of that native light which many erroneously attribute to the planets; for, by satisfactory arguments to be found elsewhere, it is quite clear that the bodies of those planets are obscure and derive their light exclusively from the Sun.

Secondly, I should be more ready to commend those who employ their skill in computing Ephemerides, if, instead of servilely receiving the report of others, they would trust something to their own eyes. Indeed no one who has eyes and who diligently avails himself of his opportunities can be said to be so destitute of astronomical instruments that he cannot observe many things in the heavens, the knowledge of which, acquired with so little trouble, would conduce greatly to the advancement of the science. And although even the best of the common tables may err, this observation alone clearly shews that there are no others which can supply their defects; nor will these tables even impugn its accuracy, as they are less at variance with it than with each other.

Thirdly, they who are so astonished at the minuteness of the diameter of Venus should rather be surprised at those astronomers whose carelessly-formed opinions have assigned such monstrous proportions to the planets; for I will prove that the diameter of Venus ought not to seem greater than we in reality have found it. But however much less it may be than the dimension usually attributed to it by astronomers, it has nevertheless far exceeded the size of any spot which I have observed. Schickard indeed remarks, that "the solar spots sometimes appear so large that they are visible through an opening in a darkened apartment; and that, from a small aperture in a wine cellar, on the 6th of July, 1629, he had observed such an one which was broader and darker than any that had come under his notice, having a peduncle in the shape of a pear." But these spots are rarely seen so large, indeed I have never yet witnessed any to be compared with this shadow of Venus, the common ones scarcely equalling half-a-minute, except when many are seen together so as to increase their bulk.

But even if this spot of ours agreed with the common ones in magnitude, yet we can shew, by other and more certain proofs, how it may be distinguished from them. I have noticed particularly three remarkable points of dissimilarity, of which the first two are probable distinctions, and the third a certain one.

First, as to figure. The figure of this body was a perfect sphere, such as is usually attributed to the planets, to the eternal bodies of the universe, and to Venus herself. But the common spots, which are nothing more than smoky exhalations, or, as one may say, solar nebulosities, consisting of fluid matter easily dispersed, are rarely found to assume a spherical form, but are of an irregular shapeless figure, and may be aptly compared with the terrestrial clouds. Moreover those spots which when seen upon the centre of the Sun appear large and spacious, when upon his limb or near the edge are compressed into a lengthened figure, and are exceedingly subtile. This proves that they do not possess a spherical or globose shape, but one extenuated and diffusely spread, and therefore that they are not stars as some imagine. Ours then is no common spot, since it retains unchanged the same spherical figure and magnitude as exactly when in the circumference of the Sun as when far distant from it.

Second, as to color. Since the ordinary spots, or solar nebulosities, are of rarer and less dense matter, scarcely darker than that of thick smoke, they cannot be said entirely to exclude the light of the Sun, but rather to transmit its rays more faintly; they are therefore seldom, if ever, perfectly black, but are more frequently a darkish kind of color mixed with light, especially round their edges which no doubt are more rare than the centre. But this beautiful shadow of Venus clearly shewed that it proceeded from an opaque and very dense body resembling the planets; for even the Moon in a solar eclipse does not cast a shadow denser, in proportion to its magnitude, than the one which I have observed from this spot.

Thirdly, and lastly. I found a remarkable difference between the motion of this shadow and that of the common spots upon the sun; so that, if other arguments were insufficient, this fact of itself proves most clearly and incontestably that it was a very unusual one, and occasioned by Venus alone. Moreover, the common spots are close to the surface of the sun and are carried round with him, performing a revolution in the space of a month, providing any of them happen to last so long. Wherefore at the beginning and end of their appearance, while passing round the receding edge of the Sun, they seem to move at so slow a rate that a day or two scarcely changes their position, their approach to or departure from our sight being as it were, in a right line. But that which we observed, passed with a rapid and uniform motion over the edge of the Sun, traversing the twentieth part of his diameter in half-an-hour, which the common spots have never done in two whole days.

Perhaps I have argued this point at greater length than it really merits; not because I thought that an astronomer would entertain a doubt as to these spots which are visible almost daily upon the Sun's disc, but that I might have an opportunity of explaining their nature and peculiarities.

For I know that there are some who make it their business to deny with the most obstinate and reckless malevolence the truth of our discoveries, and who contend that these solar spots are not temporary and fleeting vapours, but real planets and durable bodies; lest forsooth the dogma of the Peripatetics, respecting the incorruptibility of the heavens, should be impugned, which our doctrine beyond all question effectually opposes. Indeed the common spots are so different in their nature from the stars that, even in the centre of the Sun, they are frequently observed to be engendered, to increase, to diminish and to die away; of which any candid enquirer may easily convince himself. But it is in vain to speak of these things to those who will not hear, and who prefer their Aristotle, or to speak more plainly, their own unreasonable prejudices to the clearest demonstration. It is much easier to teach the ignorant than those who will not learn.

Let such men make the most of their wilful blindness, and delight in their fables; let them keep to their worthy instructor, under whose mantle they may safely retire! I envy not their ignoble dreams. At least, when astronomers meet with an observation similar to ours, let them know how to distinguish Mercury or Venus from the common spots upon the Sun.

CHAPTER V.

An Examination of the apparent Longitude and Latitude of Venus from the Sun.

A PLAIN statement of the observation having now been made, and the truth of it proved, it remains for us to explain of what advantage it may be to astronomy. In the first place, the apparent longitude and latitude of Venus from the Sun's centre are to be computed; and, with this view, we annex an estimate of the distance of their centres, and of the inclination.

But before proceeding, let us ascertain the Sun's apparent diameter; for this will be our surest guide in computing the distance between their centres. On this point astronomers differ considerably: it was according to

Kepler	31′ 1″
Tycho and Longomontanus	31′ 54″
Lansberg	35′ 50″

a very important difference, certainly, and one not easily reconcileable with the laws of astronomical science. For the present, however, I will not advert to these inaccuracies; but leave them for fuller consideration at a future time, and proceed to other matters. Let us then assume the diameter of the Sun to be 31′ 30″, which is nearly the mean of Kepler and Tycho, an estimate which I adopt, not from regard to the idle adage "medio tutissimus ibis," but because I have found it, from my own repeated observations, to be very close to the truth.

My circle having only thirty divisions, the distances before given will have to be reduced into minutes and seconds, of which the Sun's diameter will be 31′ 30″, as the following table will satisfactorily explain:

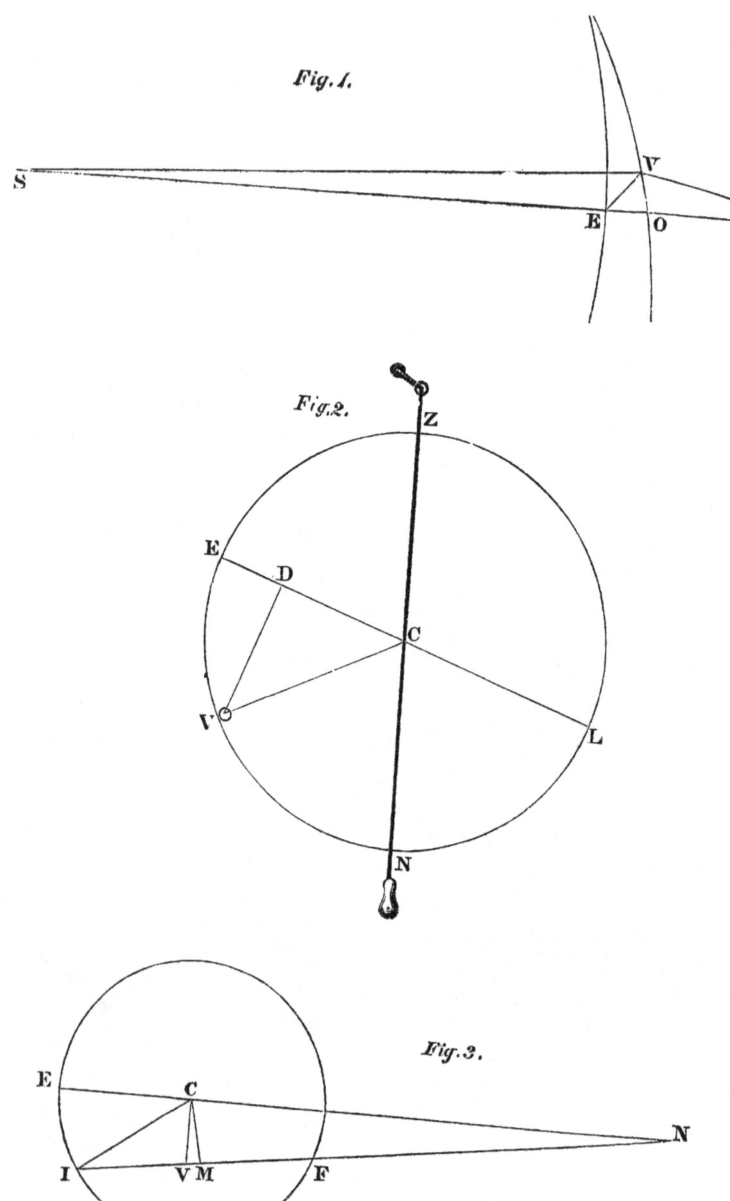

Fig. 1.
S
V
E
O
Fig. 2.
Z
E
D
C
V
L
N
Fig. 3.
E
C
N
I
V
M
F

By the Clock.	The Distances of the Centres.
3 . 15	15′ 17″
3 . 35	14′ 10″
3 . 45	13′ 39″

From these distances, together with a constant inclination of 62° 30′, the longitude and latitude of Venus from the centre of the Sun was demonstrated, as is shewn in the foregoing figure No. 2 in the plate, representing her true situation on his disc, at her first entrance.

Let C be the Sun's centre, V Venus, E C L the Ecliptic, Z C N the Vertical, Z the Zenith, N the Nadir, C V the Distance of the Centres, D C the Difference of Longitude, D V the Difference of Latitude; the angle V C N the Inclination, N C L the Parallactic Angle or the Inclination of the Ecliptic to the Vertical, E C V the Inclination of the circle through the centres to the Ecliptic.

The Parallactic Angle N C L is computed by the doctrine of spheres; the altitude of Culmination and the Sun's distance from it, together with the Meridian Angle, being given by a well-known method. To this is added the observed Inclination V C N; and thus it forms the angle V C L whose complement to a semi-circle is the inclination of a circle through the centres to the Ecliptic E C V. This being given, it will be as the radius is to the distance of the centres C V, so the line of the angle E C V is to the difference of the latitude D V: and so the sine of the complement is to the difference of the longitude D C. All of which, in the three observations, are carefully deduced in the following manner:

	D.	M.
The true situation of the sun .	12	24
The right ascension	250	55
The altitude of the Equator .	36	25

From these is given	D.	M.	D.	M.	D.	M.
The Hour	3	15	3	35	3	45
The Culminating degree	27	34	2	23	4	48
The Meridian angle	78	37	76	54	76	4
The altitude of the Culmination	15	43	16	45	17	18
The distance of the Sun from the Culmination	45	10	49	59	52	24
Therefore the angle N C L . .	70	56	68	53	67	55
To which V C N being added .	62	30	62	30	62	30
Gives the angle V C L . . .	133	26	131	23	130	25
To the complement of which E C V	46	34	48	37	49	35

Answers	M. S.	M. S.	M. S.
The distance from the centre V C	15 7	14 10	13 39
The difference of longitude D C .	10 24	9 22	8 51
The difference of latitude D V .	10 58	10 38	10 24

And thus are found the three distances of Venus from the Sun, with respect to her longitude and latitude.

In noting the observation, it is however obvious that the Inclination is uncertain to one or two degrees. Lest therefore it should be thought that any great mistake with respect to the situation of Venus might arise from this error, I will here show how little is left in doubt. Imagine then that I have erred 5°, and that the first hour of observing is 3 15′.

	D.	M.
The Inclination V C N	67	30
The angle V C L will be	138	26
To the complement of which E C V	41	34
Answers	M.	S.
The distance of the centres C V . .	15	4
The difference of longitude D C . .	11	19
The difference of latitude D V . .	10	2

The error therefore will be

In longitude	0	55
In latitude	0	56

It is clear therefore that an error of 5° in the Inclination would not alter Venus' situation, either in its longitude or latitude one minute, which is very little. But I believe that I have not erred 5°; therefore, the apparent situation of Venus being satisfactorily ascertained, I shall proceed.

CHAPTER VI.

The alteration of the apparent into the true situation of Venus.

I BEHELD Venus, during the transit, not from the centre but from the surface of the earth; therefore I observed her apparent and not her true situation. Her true situation, which chiefly concerns us, is only to be obtained by the correction of the parallaxes, into which subject I now proceed to enquire.

The hypotheses of all astronomers make the parallax of Venus in so near an approach to the earth sufficiently apparent; but this I shall leave to be further considered in a separate treatise, and in the meantime retain my own opinion.

After much and repeated consideration, I find the mean distance of the Sun to be equal at least to 15,000 semi-diameters of the earth. This paradox, as it may seem, differs greatly from the commonly received opinion; nevertheless I trust elsewhere to substantiate its correctness. Let us now ascertain, from this distance of the Sun, the distance and parallax of Venus.

According to observation, it was as the following calculation shews, chap. 14:—

The distance between the Sun and the Earth .	98409
The distance between the Sun and Venus . .	72000
Therefore the distance between the Earth and Venus	26409
Of which the mean distance of the Sun . . .	100000
But of semi-diameters this observation supposes	15000
And the distance of the Earth from Venus . .	3962

Venus therefore was distant from us just so many semi-diameters of the Earth; to which distance belongs—

	M.	S.
The horizontal parallax of Venus	0	52
From which the parallax of the Sun being subtracted	0	14
Gives the parallax of Venus from the Sun . .	0	38

Indeed so small a parallax will effect only a trifling alteration; and, if we were to take no notice of it, the inconvenience would not be much felt; but since we have leisure let us remove even these slight objections from our scrupulous opponents. It is not more trouble to apply the parallax than to investigate it.

It is a problem sufficiently well known that the parallax of the altitude of Venus, which differs nothing from the horizontal on account of the inconsiderable altitude of the Sun, is extended in length and breadth; given therefore the parallactic angle which I before computed in each of the observations, and the following parallaxes are obtained:—

The Hour.	Of the Longitude.	Of the Latitude.
3 . 15	0′ 13″	0′ 36″
3 . 35	0′ 14″	0′ 35″
3 . 45	0′ 14″	0′ 35″

Venus was with the Sun in the western quarter of the Zodiac, in longitude more east than the centre of the Sun, in latitude more south, therefore the parallax diminishes the apparent longitude from the Sun and increases the latitude; hence, in order that both may be true, we must add in the one case and subtract in the other, which being done, the true difference is given.

The Hour	Of the Longitude.	Of the Latitude.
3 . 15	10′ 37″	10′ 22″
3 . 35	9′ 36″	10′ 3″
3 . 45	9′ 5″	9′ 49″

CHAPTER VII.

An Inquiry into the Time and Place of the True Conjunction of Venus and the Sun.

I WAS not able to observe Venus at the actual point of her conjunction with the centre of the Sun, for both had set before she arrived there. But as the chief utility of the observation depends upon a knowledge of the true conjunction, I will therefore represent it from those facts which I was fortunate enough to observe.

The diurnal motion according to the calculation:

	D.	M.	S.
The direct motion of the Sun	1	1	2
The retrograde motion of Venus		36	38
Therefore that of Venus from the Sun was	1	37	40

The differences of longitude which we have found are next to be divided by this diurnal motion of Venus from the Sun, that the time may be obtained which is to be added to the moment of the observation, in order to give the true hour of the conjunction, in this manner:

	M.	S.	M.	S.	M.	S.
The difference of longitude . . .	10	37	9	36	9	5
Gives the hours.	2	36½	2	21½	2	14
Add the hour of observation . .	3	15	3	35	3	45
Which makes the hour of conjunction	5	51½	5	56½	5	59

The moment of the conjunction, which from all the observations ought to be exactly the same, shews a difference of 7½′, a small variation which the impartial reader will easily excuse. The medium between the extremes may be retained with safety, and thus ascertained will be 5 55′.

To obtain the true longitude at this moment, the Sun's situation is to be computed, the situation of Venus being apparently the same, but in reality the contrary. Therefore from my calculation—

	D.	M.	S.
The true situation of the Sun is	12	29	35
And that of Venus will be	12	29	35

So far for the longitude. But as the situation of Venus is at length clearly known, and the latitude is made evident, it is necessary to ascertain it also at the hour of conjunction.

The diurnal variation of the latitude of Venus is assumed from calculation to be 15′ 40″; and because the latitude was south around the northern node, it therefore decreased, as this observation likewise shews. The diurnal variation of the latitude must therefore be divided into the hours and minutes in which the true conjunction followed the observation, and the quotient added to the observed latitude in this manner:

	D.	M.	D.	M.	D.	M.
In hours.	2	40	2	20	2	10
The latitude decreases . . .	1	44	1	31	1	25
The observed latitude . . .	10	22	10	3	9	49
Therefore at the hour of conjunction	8	38	8	32	8	24

The first observation differs from the third 0′ 14″, which is of no importance; but if, as before, we take the mean, the latitude will be ascertained at the hour of conjunction to be 8′ 31″ south.

CHAPTER VIII.

The Demonstration of the Node of Venus.

IT will conduce much to the improvement of astronomy if the node of Venus be shewn; therefore to demonstrate this from what is already discovered, let S in the foregoing figure No. 1 in the plate represent the Sun; T the Earth; V Venus; E N the portion of the Ecliptic; O V N part of the orbit of Venus; N the Northern Node; E N V the inclination of the orbit of Venus to the Ecliptic, which on the authority of Kepler I assume to be 3° 22′; E T V the apparent angle of the latitude of Venus on the Earth 8′ 31″ from observation; S E the distance between the Sun and Venus; T E the distance between the Earth and Venus. From these the distance of the node E N from the place of the conjunction is thus computed:

1st. In the plane triangle T E V			
The right angle T E V is given	D.	M.	S.
With the angle E T V	0	8	31
And with the side T E			26409

		D.	M.	S.
Therefore the side E V		0	0	65
2nd. In the plane S E V				
The right angle S E V is given				
And the side S E				72000
With the side E V		0	0	65
Therefore the angle E S V (or the arc E V)		0	3	7
3rd. In the spherical triangle N E V				
The right angle at E is given				
The arc E V		0	3	7
And the angle E N V		3	22	0
Therefore the arc N E		0	53	10
Let the place of the conjunction be added to this	♊	12	29	35
Which makes the longitude of the node ...	♊	13	22	45
But the node of Venus is according to				
Kepler	♊	13	31	13
Longomontanus	♊	14	32	6
Lansberg	♊	11	56	4

I cannot pass over, without astonishment, this difference of opinion, so much to be regretted among astronomers of such celebrity; nor is the result unimportant, so great is the discrepancy, for it changes the latitude of Venus in this position nearly half a degree; and although elsewhere in more remote distances, the variation may not be so perceptible, yet it never disappears so completely as not to be a great reflection upon our astronomers who err to such an extent; and the more so as from other observations now extant, they might so much better agree among themselves. Lansberg, who aggravates his fault by foolish boasting, is one of those chiefly to blame; nor is Longomontanus, who possessed to so little purpose the observations of his friend Tycho, much more excusable; but here as elsewhere, the ingenious Kepler errs least of all.

CHAPTER IX.

The beginning, middle, and end of the Transit are shewn.

WE have already spoken of the hour of the true conjunction in respect of the ecliptic, but as that was not the middle of the transit, nor was there shewn in it the nearest distance of the centres, it may perhaps be agreeable to some, though it is not otherwise of much use, to assign the true middle, together with the beginning and end, of so unusual and wonderful a conjunction. For this purpose, let a figure be drawn, such as No. 3 in the preceding plate, and let C be the Sun's centre; N the Northern node; E C N the ecliptic; I N the orbit of Venus; I the beginning of the transit; M the middle; F the end; V the true conjunction in respect to the ecliptic; C V the latitude of Venus at its true conjunction; C M the least distance of the centres in the middle of the transit; C N the distance of the node from the place of the true conjunction; E N I the visible inclination of the orbit of Venus to the ecliptic. From these the periods of incidence M I and I M F are thus computed:

	D.	M.	S.
1st.—In the triangle V C N the right angle V C N is given.			
The side C N (chap. 8)	0	53	10
The side C V (chap. 7)	0	8	31
Therefore the angle C N V	9	6	0
And to this V C M is equal, whence moreover the right angle V M C is given with the side C V...	0	8	31
Therefore the side V M	0	1	21
And the side C M	0	8	24

	D.	M.	S.
2nd.—The diurnal motion of Venus from the Sun which I before used is less than in her proper orbit. To find this in the triangle V C N. Let the right angle V C N be given.			
The diurnal motion in the Ecliptic C N	1	37	40
With the angle C N V	9	6	0
Therefore the diurnal motion in her orbit V N	1	38	55
By this let V M be divided	0	1	21
The horary periods are	0	19	30
Which must be added to the moment of the true conjunction	5	55	0
That the middle of the eclipse may be found ...	6	14	30
3rd.—For the periods of incidence in the triangle I M C the right angle at M is given.			
With the side C M	0	8	24
And the sum of the semi-diameters of the Sun and Venus C I	0	16	23
Therefore the periods of incidence I M	0	14	4
Divided into the diurnal motion	1	38	55
Give the time of incidence	3	25	0
In a similar manner they are computed by the difference of the semi-diameters as in a total eclipse of the Moon.			
The periods of half the eclipse	0	12	34
The time of half the eclipse	3	3	0
Therefore the first ingress will be ... Hour	2	49	30
The total ingress	3	11	30
The middle...	6	14	30
The first egress	9	17	30
The total egress	9	39	30

CHAPTER X.

An Examination of the Calculations of Astronomers respecting the foregoing.

The value of this observation, in correcting the motion of Venus, has already been explained. We must next ascertain how the facts which are deduced from it agree with the calculations of astronomers. This inquiry will doubtless shew the usefulness of the observation to the practical student; especially as it will appear that even the best astronomers have not only disagreed among themselves, but have considerably deviated from the truth.

There are four astronomers from whose tables Ephemerides are at this time chiefly computed, into whose respective merits, as there is some difference of opinion, it may be well carefully to inquire.

1st. Copernicus who compiled the new, or rather the renewed, hypotheses, and the laws of the sidereal motions, in six books of Revolutions, from which Erasmus Reinhold afterwards constructed the Prutenic tables; and from these, Origanus, Maginus, and others derived their Ephemerides which are still extant, and are chiefly used in our prognostics, though now the Prutenic calculation is held in less esteem.

2ND. Longomontanus, the disciple of Tycho Brahé, and as it were the heir of his discoveries, who, in his Danish astronomy, treading faithfully in the footsteps of his master, brought to a conclusion those things which Tycho was by death prevented from finishing.

3RD. The sagacious Kepler, who formerly assisted Tycho in his calculations, was afterwards astronomer to three Emperors, and happily effected the renovation of the science by the publication of the Rudolphian tables, to which his other writings may be considered a prelude.

4TH. Lastly, Lansberg, who undervalued the labors of his predecessors, and with much assurance endeavoured to substitute his own perpetual tables of the celestial motions, loading them to satiety with the praises of himself and others.

I will give the calculations of these four men,

in order that it may appear who has best explained the difficulties respecting Venus, and who, in other respects, is most safely to be trusted. This observation is well suited to the purpose; for the calculation may answer tolerably well in very great distances from the Sun, though it is otherwise erroneous: greater accuracy is necessary in the inferior conjunction; and unless the calculation be, as it were, held together, it will betray gaping chinks, and the smallest error will be detected. It also happens, though why I do not know, that what ever is faulty in the hypotheses of the astronomers shews itself principally here, the errors being in this instance accumulated, and not compensating one another as is sometimes the case.

But I shall be content to set forth the calculation from their tables alone, and will not weary myself nor my readers with any geometrical delineation of hypotheses or superfluous computations of triangles; for there is no need of such nicety in refuting gross errors, neither is it necessary to waste paper in a prolix display of circles or in a description of hypotheses, which are incorrect in their very form.

Come then, ye renowned astronomers of our own times! Behold here a noble reward,—Venus promises Urania, fairer than any Helen, to him who shall happily win her.

CHAPTER XI.

The Calculations of Copernicus.

I shall commence with the incomparable Copernicus, the successful reviver of what Gellibrand calls the "noble hypothesis of the motion of the Earth," whom all the lovers of astronomy have hitherto followed, and will doubtless continue to do. Having long contemplated and admired a philosophy so sublime and so worthy of a Christian, I thus expressed my aversion to the puerile fictions of the pagan Ptolemy:—

Why should'st thou try, O Ptolemy, to pass
Thy narrow-bounded world for aught divine?
Why should thy poor machine presume to claim
A noble maker? Can a narrow space

Call for eternal hands? Will thy mansion
Suit great Jove? or can he from such a seat
Prepare his lightnings for the trembling earth?
Fair are the gods you frame forsooth! nor vain
Would be their fears if giant hands assailed them.
Such little world were well the infant sport
Of Jove in darker times; such toys in truth
His cradle might befit, nor would the work
In after years have e'er been perfected,
When harlot smiles restrained his riper powers.
These are your fancied gods, your paltry dreams;
And worthy them is all you raise around;
The temples that you build are amply large,
Thy heavens are suited to a Jove like thine.
Are such the auspices by which you rule
Your world? No longer I deplore the earth
That stands begirt with solid adamant;
Such walls repel unholy deities,
And keep the nations pure. How wisely doth it
Court repose far from the stars where it would
Have to mingle in degrading commerce,
And find, not heaven, but realms replete with crime.
Calm urge thy chariot through the starry sphere,
O Phœbus! crowds oppressed with wine can bear
No tumult. Now the banquets of the gods
Are spread by one, a youth, whose limbs betray
His steps, whose head in whirling motions lost

Can never mix the cup with steady hand.
Yet spare thyself, thy labor wisely cease,
And while the sober deities recover
Their sounder senses, let thy jaded steeds
Renew their strength with nectar and ambrosia.
No trifling task it is to hurl at once
So many gods and stars in uniform
Gyration. Then let those whose little sum
Of learning reaches but to tell the tale
Their fathers told before, whose every word
Deals in absurdities unworthy heaven,
Rival each other to applaud this fable.
But a sublimer throne is thine, and awe
Ineffable awaits thy lightning's course,
Thou God of truth whose certain laws direct
The starry spheres, whilst all the powers above
Admire and tremble; the projected Earth
Rolling along its planetary path
Hath learned to hail thy triumph; and this age
Enables mortal eyes in thy great works
To view thee nearer, and with nobler thought
To trace the stars whose order proves them thine.
In vain the Sun his fiery steeds would urge,
In vain restrain them, or attempt to guide
Their rapid course within the laws of fate.
The Earth performs their task, and by each day's
Revolving saves to all the distant stars

The useless labor of unceasing motion.
The clouds which once obscured our mental sight
Are gone for ever; great Copernicus,
Sent from above, lays open to our view
The arduous secrets of wide heaven's domain.
Turn hither then your grateful steps, for here
Are wondrous mysteries that you may learn,
Open to all whom, freed from baser thoughts,
The love of truth impels, and whom no cry
Of vulgar men can scare from what is right,
Nor fear oppress, O child of ignorance!
Nor fabling oracles once deemed divine.

It was sufficient for Copernicus to have laid so good a foundation, we must pardon him if, his sublime understanding being perplexed by some few inaccurate and fallacious observations, he failed in rearing the superstructure; for he neither discovered the true form of motion, nor did he ascertain the numbers with precision, being too much devoted to the circles and equality of the ancients, as appears from this observation which I thus calculate from his tables, assuming the difference between the meridian of Frueburg and our own to be 1° 30′.

Of the Sun.	sex.	deg.	min.	sec.
Simple equable motion (æqualis simplex) .	3	44	14	29
The simple anomaly of the Equinoxes . .	2	58	40	46
The prosthaphœresis of the centre to be added		0	10	53
The proportional parts (scrupula proportionalia)			0	0
The mean anomaly of the Sun	2	31	53	16
The coequate anomaly (anomalia coequata) .	2	32	4	9
The prosthaphœresis of the orbit to be subtracted		0	53	12
Therefore the true simple motion of the Sun	3	43	21	17
Of Venus				
The apogee	0	48	20	0
The anomaly of the centre	2	55	54	29
The prosthaphœresis of the centre to be subtracted		0	8	43
The proportional parts (scrupula proportionalia)			59	53
The eccentric longitude	3	44	5	46
The mean anomaly of the orbit	2	58	48	7
The equate anomaly of the orbit	2	58	56	50
The prosthaphœresis of the orbit to be added		2	50	20
Therefore the situation of Venus by the fixed stars	3	46	56	6
The south latitude		0	21	30

In the latitude there is a small error, not indeed more than 13′; but in the longitude there is a very considerable one, for Venus, who was actually in conjunction with the Sun, was distant from it, according to this calculation, 3° 34′ 49″, and as her diurnal motion from the Sun is 1° 37′ 40″, they were in conjunction the day after, at four minutes and forty-seven seconds past two.

Therefore it is not on account of Mercury alone that Schickard may pity the vanity and unskilfulness of the astrologers who, putting forward their tables as true, trifle with the fate of posterity. Venus does not smile upon their absurdities: what good luck is destined for me? what sort of a wife? the inconstant Mercury is propitious, will not Venus, whom the astrologers conciliate by such well-contrived calculations, be so likewise? I perceive that I must apply for other assistance than the scheme of my nativity affords which, so far from telling my fortune, does not even indicate what is already revealed. Are the astrologers then, who are so profoundly ignorant in certainties, to be credited in doubtful matters?

I have computed the situations of Venus and the Sun from the fixed stars, because we are here seeking their distances only; but if you should desire the longitude from the true equinox, add to their situation, with reference to the fixed stars, the true precession of the equinoxes 28° 27′ 23″, and you will obtain it.

CHAPTER XII.

The Calculation of Lansberg.

LANSBERG, a true disciple of Copernicus, follows him very closely; indeed his numbers only differ slightly respecting some of the planets; but his formula of the hypotheses scarcely varies from that of his master. His astronomy is therefore nothing more than a second edition of the Prutenic tables. In some things perhaps he is a little more elaborate; but, in most, certainly more faulty than his original. Nevertheless he earnestly recommends his immortal fame to posterity; and, under a pompous title, offers his tables as compiled from and agreeing with all sorts of observations, without fear of detection. Let him not be angry if we should prefer, rather than himself, those whom he so superciliously condemns: and that it may be known with what justice he so confidently boasts of his own labors, let him explain, in his own words, that most accurate calculation which he has made the subject of so many encomiums.

From the commencement of the Christian era to the time of this observation there are 1638 full Julian years, 10 months, 23 days, 5 hours, and 55 minutes, under the meridian of Liverpool; under that of Goesa 6 hours and 20 minutes apparent time, or when properly corrected 6 hours and 4 minutes, this is, in *Sexagenæ dierum*, 2‴ 46″ 16′, 46 days, 15′ 10″,* by which the following motions are given.

OF THE EQUINOXES.	SEX.	DEG.	MIN.	SEC
The anomaly	5	58	32	51
The prosthaphæresis to be added			12	30
OF THE SUN.				
The mean motion (motus medius)	4	13	3	38

* As the general reader may not understand this mode of calculation, it may be well to state that Horrox takes it from Lansberg who adopts, for the arrangement of his tables, what he called *Sexagenæ dierum*. According to his method, 60 days make a *sexagena prima*, 60 times 60 or 3600 days a *sexagena secunda*, and so on. Hence, in conformity with a calculation which he gives, we have

	3a	2a	1a	Dies
1600 Julian years	2	42	20	0
38 do. do		3	51	19
The ten first months of a common year (1639 was not bissextile), or $\frac{304}{60}$ . .			5	4
Additional days in November				23
	2	46	16	46 of

time calculated in *sexagenæ* ascending; together with 15′ 10″ of *scrupula* descending.

	SEX.	DEG.	MIN.	SEC.
The anomaly of the centre	3	16	48	7
The prosthaphæresis of the centre to be added		1	42	50
The proportional parts (scrupula proportionalia)			1	20
The mean motion of the apogee	1	35	54	49
The equate motion of the apogee	1	37	37	39
The true anomaly of the orbit	2	35	25	59
The prosthaphæresis of the orbit to be subtracted		0	51	47
The mean motion of the Sun from the true Equinox	4	13	16	8
Therefore the Sun was in	4	12	24	21
Of Venus.				
The mean motion of the apogee	1	31	47	11
The anomaly of the centre	2	41	16	27
The prosthaphæresis of the centre to be subtracted		0	39	9
The proportional parts (scrupula proportionalia)			58	12
The longitude of the centre	4	12	24	29
The mean anomaly of the orbit	2	59	50	31
The equate anomaly of the orbit	3	0	29	40
The prosthaphæresis of the orbit to be subtracted		1	19	52
Therefore the longitude of Venus from the mean Equinox	4	11	4	37
From the true Equinox	4	11	17	7
The mean motion of the northern node . .	1	11	43	34
The distance of Venus from the northern node	3	0	40	55
Therefore the north latitude of Venus . .		0	10	45

The observation shews Venus in conjunction with the Sun; this calculation separates them 1° 7′ 14″.

Therefore the conjunction by computation was earlier by 16 hours 31 minutes.

The observation decreases the latitude south, while the calculation increases it as much north. Hence the studious may perceive how little these perpetual tables, which their author so loudly praises, are to be relied upon; certainly a little more modesty would have been more consistent with their pretentions than so many undeserved compliments, which among prudent people have the effect of lessening rather than of increasing confidence.

No one who is disposed to favor Lansberg must be blamed; the diameter and parallax are, in his opinion, assumed to be different from these statements. But, if we should follow him in the longitude, both causes, and in the latitude the former, would increase the error.

CHAPTER XIII.

The Calculation of Longomontanus.

It may perhaps be some consolation to the admirers of the Lansbergian astronomy, if there are any, to learn that the followers of Tycho, disowned by their master and to whom Hortensius, the advocate of Lansberg, strenuously denies the merit of having perfectly restored the science of astronomy, (see Preface to Lansberg's Motion of the Earth), labor under a similar or even a greater error; and, lest I should seem to envy them the miserable satisfaction "habuisse socios," I will edify their dull souls by convicting Longomontanus, Tycho's disciple and his too faithful follower in all things whether true or false, of a most palpable blunder. His calculation is thus:

To the current year of our Lord 1639, 24th day of November 5 hours 55 minutes at Liverpool; or 6 hours 52 minutes by apparent time, and 6 hours 46 minutes by mean time at Uraniburg these motions are given.

	SEX.	DEG.	MIN.	SEC.
Of the Equinoxes.				
The anomaly	3	20	30	28
The prosthaphæresis to be added			9	36
Of the Sun.				
The equable motion (motus æqualis) . . .	4	13	9	13
The apogee	1	36	15	14
The anomaly of the orbit	2	36	53	59
The prosthaphæresis of the orbit to be subtracted		0	49	46
The mean motion from the true equinox . .	4	13	18	49
Therefore the Sun's situation	♐	12	29	3
Of Venus.				
The apogee	1	30	22	30
The anomaly of the eccentric	2	42	46	43
The prosthaphæresis of the eccentric to be subtracted		0	33	5
The proportional parts (scrupula proportionalia)			58	30
The eccentric longitude	4	12	36	8
The mean anomaly of the orbit	3	0	20	55
The equate anomaly of the orbit	3	0	54	0
The prosthaphæresis of the orbit to be subtracted		2	28	37
Therefore the longitude of Venus from the mean equinox	4	10	7	31
From the true equinox	♐	10	17	7
The mean motion of the northern node . .	1	14	22	30
The distance of Venus from the northern node	2	58	13	38
Therefore the south latitude		0	7	40

The latitude is sufficiently correct, but the longitude errs 2° 11′ 56″, and hence it is one day, eight hours, and twenty-five minutes too little. In the latitude, therefore, he is more correct than Lansberg, but in the longitude he is almost twice as much at fault; nevertheless I do not wish it to be thought, from this one instance, that Lansberg's tables are superior to his in other matters, for I have often proved that Longomontanus is more correct as to the three superior planets, and also with respect to the moon.

CHAPTER XIV.

The Calculation of Kepler.

But I leave these patrons of circles and equality, these artificers of an useless labyrinth, and their hypotheses which are faulty in their construction and incapable of amendment. For although the measures of the eccentricities of the orbits, together with the mean motions, might be corrected so as to resemble this and other observations; yet as the stars are governed by different laws from those which they have invented, it is impossible by a complication of such circles to bring about an entire agreement with appearances.

I hasten therefore to that prince of astronomers, Kepler, to whose discoveries alone, all who understand the science will allow that we owe more than to those of any other person. I venerate with the greatest honour and admiration his sublime and enviably happy genius; and if necessary, I would defend with my best efforts the Uranian citadel of the noble hero who has so much surpassed his fellows, nor shall any one while I live, violate his ashes with impunity. His death was an event that must ever have happened too soon; the science of astronomy received the lamentable intelligence whilst left in the hands of a few trifling professors who had kept themselves concealed like owls until the brightness of his sun had set.

Who, mighty shade, shall sing thy praises? who,
Worthy so great a task, shall reach the stars?

Who now shall chant thy fate? The modern seers
Portend that heaven's disturbed by monsters which
Are unintelligible to mankind;
Perchance in pity thou dost still protect
The weaker minds of those whom thy decease
Hath robbed of nature's best interpreter.
Since such a guide is lost, what other now,
Deserving to succeed, can take the reins?
Or should the stars rebel, who can restore
Them to their course, and bind with closer ties
Their wandering ways? O! thou alone couldst take
The arduous guidance and shake the strong rein
To urge along the slothful retinue;
By thee restrained, the vulgar crowd
Dared not to climb the sacred car of heaven.
No devious course could cause thy thoughts to wander
In perplexity; fictitious circles
Could not enthrall thy loftier genius;
But thy mind, intent on the sublime, with
Faithful hand traced the motions which the God
Of nature hath decreed. While yet the power
Was thine to guide their way, true to thy rules
Each planet in its ordered path revolved,
And all rejoiced to follow in thy train.
But now deprived of thee science declines,
Sinking in antiquated errors; all
The stars are hurled as madness may devise,

And heaven's deformed by senseless violence!
Unhappy Germany! though torn by wars,
The sword alone will not effect thy ruin;
A heavier curse conspires to bring about
Thy mind's destruction. 'Tis this encourages
Hortensius to insult Pelides' dust;
By this the pompous Belgian, bolder grown,
Imposes on the world Perpetual Tables,
And spurns the embers which a powerful flame
Has sadly left; nor does he even fear
Lest his bold thefts should haply be detected,
Now that great Kepler's numbered with the dead.
Chaos is come again, the world's unhinged,
All things, in thee o'erpowered by fate, betray
The noblest art to trifling sycophants.

Kepler's Rudolphian tables give the following calculation of the observation, the time having been before reduced and settled by Longomontanus.

Of the Sun.	sex.	deg.	min.	sec.
The equable motion (motus æqualis)	4	13	18	7
The apogee	1	36	24	5
The mean anomaly	2	36	54	2
The equation to be subtracted		0	49	32
Therefore the Sun's situation	♐	12	28	35
The distance between the Earth and the Sun				98350

Of Venus.	SEX.	DEG.	MIN.	SEC.
The equable motion	1	13	19	2
The aphelion	5	2	4	57
The mean anomaly	2	11	14	5
The equation to be subtracted . . , . .	2	10	36	4
Therefore the eccentric longitude	1	12	42	58
Reduced to the ecliptic . , ...	1	12	43	4
The distance between the Sun and Venus .			72084	
The anomaly of the commutation	3	0	14	29
The prosthaphœresis of the orbit to be subtracted		0	39	43
Therefore the apparent situation of Venus	♐	11	48	52
The northern node	1	13	31	13
The distance of Venus from the northern node	5	59	11	45
Therefore the south latitude		0	7	45

In the longitude there is an error of 39′ 43″, which is as much as the prosthaphœresis of the orbit, and gives 9 hours 46 minutes, by which quantity the conjunction was earlier.

In the latitude, the calculation is only slightly defective. Hence it is clear that Kepler's tables represent the situation of Venus in the Sun the most correctly of all, and in this respect at least, are to be preferred. I have also found them better in various ways, both from my own observations and from those of others.

CHAPTER XV.

Correction of the Motions according to Rudolphi.

SINCE the error which I discovered in the Rudolphian tables is so great, it may not be amiss to shew how the calculation may be amended in order to agree with this and other observations. I quite agree in the form of Kepler's hypotheses, and gladly receive both his annual and diurnal motion of the earth. I am of opinion also that these motions do not arise from complicated fictions of useless circles, but from natural and magnetic causes, and that they are owing to the rotation of the Sun on its axis. He knows but little of astronomy who is ignorant that the figure of the orbit is elliptical; that its centre is the body of the Sun, and not a fictitious point near it: that the motion of the planet is really unequal; that the whole apparent inequality does not proceed from its eccentricity alone; and finally, that the inclination of all the orbits to the ecliptic is not influenced by the annual motion, but is fixed and constant. No one, we repeat, who denies such facts is sufficiently acquainted with astronomical observations. They are all fully demonstrated by Kepler, and I have found them, by subsequent examination, to be strictly true; but with the view of attaining greater perfection in the theory constructed upon these principles and in the quantity of the mean motions and eccentricities of the orbits, I have attempted to correct the motions of the Sun and Venus in the following manner; an undertaking which could not be displeasing to Kepler himself, as he frankly confessed that these matters were not yet thoroughly explored.

I. OF THE SUN.

1. The mean motion of the Sun, as to its periodical quantity, is correctly determined by Kepler, but it seems to me that one minute should be subtracted from its roots; the places of the fixed stars however ought not on that account to be diminished, as Longomontanus has hastily concluded.

2. The apogee is right in all respects.

3. The eccentricity which he makes 1800 with a radius of 100,000, I make, for many reasons, only 1735. Therefore the greatest equation will

be, according to me, 1° 59′ 18″; whereas according to him it is 2° 3′ 46″; and herein lies Kepler's principal error which has betrayed him into many others, as I shall shew at another opportunity.

4. The last correction which I shall make relates to the triple method of equalizing the natural days in the astronomical or Emperic demonstration of Tycho, and in the physical one of Kepler. The correction of the lunar motion requires this, and the diminished eccentricity of the Sun explains the difficulty in which Kepler was so deeply involved; but more of this in its proper place, God willing.

II. OF VENUS.

1. I find the mean motion of Venus much slower than Kepler makes it, namely about 18′ in a hundred years; but in the beginning of the present year, 1640, 9′ 20″ should be subtracted, and hence arises the chief cause of the great discrepancy in the calculation of Rudolphi concerning this observation.

2. The aphelion, in this age, remains at 5° in ♒; and the observations of our predecessors seem to allow it scarcely any, or at least, an ex-

ceedingly slow motion. Hence it is clear why those who refer the eccentricities of the planets to the centre of the great orbit of the Earth, find the eccentricity of Venus less at this day than what Ptolemy has recorded; for he added, during the advance of the apogee, the moveable centre of the orbit of the Earth to the fixed centre of the orbit of Venus.

3. The true eccentricity is 750, and the semi-diameter of the eccentric of Venus 100,000; therefore its greatest equation is 51′ 34″, whereas according to Kepler, the former is 692, and the latter 47′ 36″.

4. The radius of the orbit of Venus is to the orbit of the Earth as 72,333, not 72,414 as he fixed it, to 100,000.

5. It has already been demonstrated that 8′ 30″ are to be subtracted from the northern node, from the beginning of the year 1640, which may also be done hereafter in other ages.

6. The inclination of the orbit to the ecliptic appears slightly to exceed the calculation of Kepler. He has fixed it at 3° 22′ whilst I make it 3° 24′; but certainly it is not so much as 3° 30′, as Lansberg and Longomontanus suppose.

I partly began these corrections of the Rudolphian tables before the transit of Venus, from other observations; and afterwards considerably amended them by further experiments very carefully instituted. I have also brought this calculation, otherwise tolerably exact, to coincide even in the minutest particulars with our observation, in the following manner:—

Of the Sun.	SEX.	DEG.	MIN.	SEC
The equable motion (motus æqualis)	4	13	17	22
The apogee	1	36	24	5
The mean anomaly	2	36	35	17
The equation to be subtracted			47	47
Therefore the situation of the Sun	♐	12	29	35
The distance between the Sun and the Earth				98409
Of Venus.				
The equable motion (motus æqualis)	1	13	10	16
The aphelion	5	5	0	0
The mean anomaly	2	8	10	16
The equation to be subtracted			40	47
Therefore the eccentric longitude	1	12	29	29
Reduced to the ecliptic	1	12	29	35
The distance between the Sun and Venus .				72000
The northern node	1	13	22	45
Distance of Venus from the northern node .	5	59	6	44
Therefore the south latitude			8	31

You see here that, agreeably to our expectation, Venus was exactly conjoined with the centre of the Sun; therefore there is no anomaly of the commutation, nor prosthaphœresis of the orbit. You also see that the latitude and other particulars exactly agree with the observation; this result indeed might easily be obtained from a single example, but it would be tedious, and foreign to the subject in hand, to shew what might happen in other circumstances. I ask therefore that credit may be given to my bare word for the present; and, with God's permission, by further collating and condensing my proofs, I will cause Venus to arise from this sea of error, to come forth, wrapt in the chain of numbers, more beautifully than she did from the arms of Vulcan, and to learn a modesty unprecedented in her former deportment; nor, as heretofore, shall she wander in wanton lasciviousness, evading and despising the care of her guardians whose councils have been so little attended to, as we have already plainly seen:

Tantæ molis erat muliebrem frangere mentem.

CHAPTER XVI.

On the diameter of Venus.

CONGRATULATE us, Gassendi, on clearing from suspicion your observation of Mercury, and let astronomers cease to wonder at the surprising smallness of the least of the planets, now they find that the one which seemed the largest and brightest scarcely exceeds it. Mercury may well bear his loss since Venus sustains a greater.

I observed the diameter of Venus (Chap. I.) to be 1′ 12″, the Sun being 30′; therefore the latter being 31′ 30″, the true diameter of the former is 1′ 16″. My friend Mr. Crabtree's observation agrees with this calculation: I am sure she did not appear greater; if there is any error, it is in an excess. There is no reason why any one should doubt the truth of the observation; unless indeed he is unacquainted with the telescope, or influenced by the knavery of the Peripatetics, or suspects our honesty; and I all not stay to argue either with those who have not seen this instrument or who mistrust its fidelity, for it is vain to contend with ignorance and self-will. Permit me to remind any who may suspect our good faith, how easy it would be to investigate the subject for themselves, and how little it would serve our purpose to distort truth by falsehood.

Let us then examine the opinions of others, in order that it may appear with what degree of accuracy astronomers have hitherto estimated the magnitudes of the stars.

1. Tycho Brahé, in whom most men place confidence in such matters, makes the diameter of Venus 3′ 15″ in her mean distance from the Earth. But the distance of Venus from the Earth according to our observation was 26,409, and the mean distance of Venus or the Sun from the Earth 100,000 as was before shewn; therefore Venus, who from the distance of 100,000 appears to be 3′ 15″, at the distance of only 26,409 will be 12′ 18″. But this is far from the truth, being nearly ten times as much as in the observation.

2. Philip Lansberg, who boasts so authoritatively of his Uranometria, makes the diameter of Venus in her mean distance 3′ 0″; therefore at the distance before-mentioned, it would be

11′ 21″. This is very far from the mark, being nine times greater than in truth it should be.

3. From the tables of Rudolphi, according to the precepts of Kepler, the diameter of Venus, by our observation, is computed to be 6′ 51″; his is the nearest approach to the truth, as is generally the case with Kepler, but still it is five times or more in excess.

Copernicus and Longomontanus say nothing of the diameters of the five primary planets; but the ancients, Alphraganus and Albategnius, differ very little from Tycho and Lansberg.

Since therefore the observed diameter of Venus differed so considerably from what has been assigned by the whole school of astronomy, it may perchance be doubted whether some optical deception has not caused it to appear small; for Schickard, an excellent mathematician and professor of Hebrew and astronomy in the university of Tubingen, supposed that such was the case with respect to the Mercury of Gassendi, the minuteness of which caused equal astonishment. The reasons why he supposed Mercury in the Sun to be diminished below the truth, as they apply equally to Venus, I shall briefly subjoin, and with the author's permission, examine; for I observe that some sensible men acquiesce in his opinion, and, from not having sufficiently considered the subject, at once take for granted that which connects, upon any grounds, new appearances with old opinions.

1. He takes his first argument from the diffusion of the solar light. "You know" says he "it is the nature of this light to spread and diffuse itself on all sides, hence it necessarily follows that opaque bodies in the immediate neighbourhood are somewhat divided and cut away. You may see this in a familiar experiment which I have often tried by candle-light among my winter amusements; if you cause a short stick to be held out at a short distance, you will find that as you stand apart from it, it will appear to be serrated on both sides where the light crosses it, as if it were cut and ragged."

2. He argues from the opticians Alhazen the Arabian, and Vitellio, the Sarmatian, who shew that the base of the shadow is less than the hemisphere of its body, if the illuminating sphere

be greater than that which is illuminated; whence he assumes as certain that "nothing could be seen of Mercury or Venus in the Sun, except what was turned away from its light and placed in the shade; and that this must be less than half, since the illuminated part is greater than half; therefore Mercury, and consequently Venus, appear to be small."

3. He gives another reason which he confesses to be only probable: "If it be right to reason from the analogy of the moon to other planets, we must believe that they are not all obscure, but have opaque parts in the middle, or nuclei, whilst externally they are covered with a kind of transparent coating like a mirror, the one part representing the metallic foil and the other the glass which reflects the rays that fall upon it; for when the moon approaches the stars, she seems to envelope them as they draw near and to admit them somewhat within her luminous periphery; on the contrary, when they are receding, she seems to restore them to sight before they touch her border. Mœstlinus noticed this in the cases of Mars and of the heart of the Scorpion, in the

year 1595 (*Disput. de pass. plan. Thes.* 148) whence he inferred that they are surrounded by a kind of transparent air. But I leave this for more mature experience."

With your leave, most learned Schickard, I must entirely differ from you in this particular, for I do not believe that either your Mercury or our Venus were at all less than the true measurement requires; nor are they in the heavens different from what they appear to us in the Sun, unless that the radiations might interfere and increase their visible magnitude in the day time, though this would not affect bodies seen upon the Sun's disc. You will therefore allow me to prefer the simple truth to your arguments which I think may be easily confuted.

1. I readily admit that there is a remarkable, and indeed an almost incredible, diffusion of light when we gaze upon it with the naked eye; and I wish that astronomers would sufficiently bear this in mind, and that they would not allow the false rays of the planets and fixed stars to deceive them by making the true magnitude of Venus and Mercury seen in the Sun to appear so astonishing

owing to this delusion. Contiguous opaque bodies are certainly divided and cut away, when beheld by the naked eye, but not otherwise: but your experiment of the stick seen in the candle-light, although it may be true, does not appear to have any reference to the point at issue: for the reason why the light of the candle diminishes the magnitude of the stick is because its rays are refracted and amplified by the moisture of the beholder's eye; but if you look upon the shadow of the stick upon the wall it will not be at all less than the stick itself, unless the light be larger than the object and the shadow be diminished at a certain distance according to a geometrical law. But we observed the shadows of Mercury and Venus depicted in the light of the Sun, through the telescope by which the rays are so modified as to be easily endured by the eyes. Indeed if we had tried to observe the planets in the Sun with the naked eye, I can readily conceive that we should not have been able to see them at all; for the diminutive bodies of Mercury and Venus would have been entirely concealed from our view, owing to the powerful light of the Sun being so oppressive. But in a darkened view, the affair is very different; and there is no reason to fear the light of the Sun diffusing itself more than is legitimate or cutting off the contiguous opaque bodies beyond what is proportionate.

We have a much better experiment when the moon eclipses the Sun. The naked eye always estimates the eclipse less than the truth, as may be proved by many examples; but the telescope exhibits the exact quantity, both of the eclipse and of the lunar diameter. I lately proved this in the eclipse of the Sun on the 22nd of May 1639; and Gasendi observed the same thing in a similar eclipse on the 11th of May 1621, when the diameter of the moon appeared by no means less than as observed at other times. Although the moon when at her full seems to be enlarged beyond her proper size, yet this is a deception which does not occur in an eclipse of the Sun. Moreover you yourself know the absurdity of the dogma for reducing the semi-diameter of the new moons, which Tycho, and after him Longomontanus sought to put upon us. Why then, let me ask, do you maintain that so zealously in Mercury which you properly reject as untenable in relation to the moon?

2. Let it be conceded to you that the Sun illuminates more than half of the bodies of Mercury and Venus, and hence, since those bodies are precisely spherical, that they are less than half in the shade: now in your turn you must allow that that which, on this account, is slightly diminished, is diminished still further from a prior cause which deceives the eye in a most remarkable manner. The amount is indeed so small that it is scarcely worth naming; but, lest the uninformed should be misled, I will explain how it arises: The diameter of the Sun, as seen from the Earth, at the distance of 98,409 parts, appeared to be 31′ 30″, and from Venus at the distance of 72,000 to be 43′ 3″; but the diameter of Venus from the Sun appears 0′ 28″, therefore the angle of the cone of the shadow of Venus will be 42′ 35″, which, being subtracted from the semi-circle, leaves the circumference of the shadow 179° 17′ 25″, the half of which 89° 38′ 42½″, 999,980,820, is the sine to the radius 1,000,000,000, and the apparent diameter of Venus is 1′ 16″ to

the true which is 1′ 16″ 0‴ 5⁗. But after all of what consequence is a trifling difference which does not exceed 5⁗? Or how can the prior cause, which is of itself of no importance, be deemed to increase a discrepancy which is so small?

But since it pleases you to debate so ingeniously, I will reply with a similar subtlety. I deny that the Sun illuminates more than one half, or that the planet appears less to us from any such reason; on the contrary, he illuminates less than the half, and so far are we from seeing the illuminated portion of the hemisphere, that we cannot discern the whole of that which is obscure, the dark part being greater than the portion which is irradiated: for I have no doubt that the bodies of all the planets, and especially of Venus on account of her strong reflection, are mountainous and uneven like the moon and the Earth. These mountains therefore will obstruct the rays of the Sun so that they cannot extend beyond the half; indeed they will not reach over more than the half of the mountains which intervene on every side, and obstruct the rays of light towards the even ground.

This is the case as regards the Earth where the Sun frequently conceals himself behind the mountains before he reaches his true setting; and these mountains terminate our view so that it does not extend as far as to the middle; accordingly the apparent magnitude would be increased rather than diminished thereby. But these are trifles.

3. What you advance in the third place is by no means proved, nor do you certainly state, such is your modesty, that the light of the Sun is reflected from the moon and planets as from a looking-glass. The idea is less common than ridiculous, for the least part of a spherical glass reflects the light of the Sun, though all surrounding objects should remain in obscurity. It is true that, on account of its great distance, the particle cannot be seen, but if it could, it would appear to be circular like the Sun; for the same reason, the moon never appears forked; indeed the object would become invisible. See a dissertation on this question by that acute astronomer Galileo in his *Cosmic System*.

Moreover the lunar mountains seen through the telescope plainly shew, from the very dark shadow which they cast, that the external surface of the moon is not transparent. Hence it is evident that her exterior matter is not less opaque than that of our Earth: nor do you consider that to entertain a contrary opinion is tacitly to confirm the Tychonian diminution of the moon in solar eclipses, which you elsewhere condemn as absurd.

I have not the least doubt but that the moon is surrounded by a kind of transparent air; nor do I think otherwise of the rest of the planets whose radiation is, on that account, very likely to be augmented. For the same reason the moon may seem to envelope the stars before they actually reach her edge, especially if she be seen with the naked eye, and the star is in contact with her lucid margin; but if you view her with the telescope covering the stars with a dark shade, you will perceive that, as they approach her edge, they very suddenly vanish. William Crabtree and I observed this most clearly in the conjunction of the moon and Pleiades on the evening of the 19th of March in the year 1637. These circumstances therefore do not by any means increase the magnitudes of Venus or Mercury.

Although Mercury rising from the horizon at Aix in Provence, together with Arcturus, on the 10th of October 1621, appeared equal to it in the eyes of Gassendi, yet this is no disparagement to the observation of the transit. For albeit that star is commonly estimated 2′, it is nevertheless very properly taken by you to be much less than 1′. Galileo found, by a singular method of observation, that the diameter of a fixed star of the first magnitude was not greater than 5″; and if the fixed stars did not shine by their own light, they would perhaps appear to be much less: the telescope, by which they are so much more distinctly seen, represents them as mere points, as was evident in the conjunction of the moon with the Pleiades; for as soon as the moon covered the bodies of the stars, their false rays immediately vanished, whereas if these had proceeded from the bodies of the stars themselves, they would have subsided gradually and not suddenly.

I greatly wonder that all astronomers should have been so much deceived in computing the diameters of the planets which they make five or six, and in some instances even nine or ten times as great as they ought to be. I think however that I understand the cause of the error, which is that they have not taken these adventitious rays into consideration. Still it surprises me that they should all have been so negligent as not to perceive a deception so remarkable as to be detected even by the naked eye. For I have often observed both Venus and Jupiter, during the day, when the Sun's altitude was some degrees, to be so minute that they could scarcely be discerned, and I have, in imagination, compared their diameters with those of the Sun and moon; but they seemed to defy all computation, and not to equal one-hundredth part of the diameter of the former luminary, whereas the common opinion supposes them to be a tenth or even a sixth or fifth. Galileo notices this error in estimating the diameters of the planets and fixed stars, and gives a method of measuring them even without the aid of a telescope, which I have frequently tried with respect to Venus, and by which, although I may not have ascertained the truth very accurately, I have discovered the greatness of the common error.

On the 7th of January in the present year 1640, the Sun being risen and diminishing the rays of Venus by his own light, an iron needle whose diameter was 8 parts at a distance of 4300 covered the planet Venus; therefore the diameter was 0′ 38″.

On the 29th of January in the same year, a needle of 5 parts covered Venus at the distance of 383; therefore the diameter was 0′ 27″.

In these observations I looked through a small opening made with a fine needle in a piece of card; by which method alone, even on a dark night, the diameters of the planets appear to be wonderfully reduced: so that, unless you are very strong-sighted, you can scarcely discover either the planets or the fixed stars which deceive the naked eye from their rays being so entirely cut off by the narrow opening.

For these reasons, I have no doubt that the diameter of Venus in the Sun appeared its proper size, and did not differ one second from the truth.

CHAPTER XVII.

Of the Diameters of the rest of the Planets, of the Proportion of the Celestial Spheres, and of the Parallax of the Sun.

I SHALL here say something which may tend to throw light upon the dimensions of the stars, and upon the horizontal parallax of the Sun, a matter of the greatest importance, and one which has been the subject of much fruitless speculation; but I will not speak dogmatically, nor, as I may say, "ex cathedrâ," but rather for the sake of promoting discussion, and with the view of examining other men's opinions.

John Kepler, the prince of astronomers, speaking of the relative proportion of the planets (*Astr. Cop.* page 484), thinks it "quite agreeable to nature that the order of their magnitudes and of their spheres should be the same; that is to say, that of the six primary planets, Mercury should be the least, and Saturn the largest, inasmuch as the former moves in the smallest, and the latter in the largest orbit."

"But as the dimensions of their bodies may be regarded as threefold, either according to their diameters, their superficies, or their bulk," he is doubtful which should be preferred. He thinks the first proportion "to be beyond question contrary to original reasons, as well as to the observations made on the diameters by means of the Belgian telescope." He advocates the second, because the original reasons are preferable; whilst Remus Quietanus, a man well versed in practical observations, defends the third; and with him Kepler at length agrees, retaining this proportion in the Rudolphian tables. But as this was not found to be entirely satisfactory, he sought a proportion in the density of the matter, whereby the bodies of equal magnitude may differ in weight, and vice versâ.

To give my opinion upon the subject, I am persuaded that the proportion of the globes and orbits of the planets is the most accurate and certain, for such would appear the most agreeable to the Divine Nature which formed all things by weight and measurement, and as Plato says, "æternam exercet geometriam." Moreover the proportion that obtains between the periods of the motions of the planets and the semi-diameters of the orbits is most exact, as Kepler, who discovered it, very justly remarks, and as I have accurately proved by repeated observation. Indeed there is not an error even of a single second. Since therefore it is true that the Sun by its attractive power regulates the motions of the six primary planets, I cannot conceive how it could adapt that power so perfectly to their several distances, unless those moveable globes themselves were similarly proportioned. In short, a well-conducted inspection of the diameters clearly warrants the same conclusion ; neither is it necessary with Kepler to have recourse to material density.

What then, you will ask, is the proportion of these orbits and bodies ? I reply, that it is the first one which has reference to the diameters, and which Kepler and others very inconsiderately reject ; and this proportion is more acceptable from its suitableness, and has been more corroborated by my own observation than that of either superficies or bulk.

For what, I ask, can be more absurd than to compare the semi-diameter of the orbit with the superficies or magnitude of the planet, rather than with its semi-diameter? It is as though we were to compare the head of one person with the foot of another, or as the poet says:—

> "Humano capiti cervicem pictor equinam
> Jungere si velit, et varias inducere plumas
> Undique collatis membris."

But on the other hand, what can be more appropriate than that the diameters of the orbit and of the planet should be proportioned to one another? According to this relation, both their superficies and magnitudes should be similarly proportioned. If Peter be twice as tall (*altior*) as John, it is not necessary in order to preserve the proportion, that his head be twice as great, (*majus*) nor twice the superficies, but twice as thick (*crassius*); and the matter will stand thus: as the body of Peter is to the body of John, so is the head of Peter to the head of John, in whatever proportion, whether of thickness, (*crassitudinis*,) or of superficies or bulk (*corpulantiæ*);

and so it is with regard to the spheres. For, because Saturn is nearly ten times taller (*altior*) than the Earth, he will not therefore be ten times greater, (*major*,) nor have a superficies ten times larger; but inasmuch as they are spheres, the orbital diameter of Saturn will contain ten times that of the Earth. Indeed any proportion may be calculated in this manner; for as the diameter, superficies, or bulk of the sphere of Saturn is to the diameter, superficies, or bulk of the sphere of the Earth, so is the diameter, superficies, or bulk of the globe of Saturn to the diameter, superficies, or bulk of the globe of the Earth; and so it is with regard to the rest.

But let us pass on to notice the observations upon which they chiefly rely who reject these arbitrary proportions as vain. It is clear from the example of Venus that experience is entirely against the proportion of Kepler; and this is also evident from Gassendi's observation of the planet Mercury, the diameter of which he found to be scarcely equal to the third part of a minute, although Kepler's calculation extends it to three minutes. The same is the case with reference to Mars whose diameter, according to Kepler's rules, is sometimes increased beyond six minutes; whereas, in reality, it never equalled two: and Kepler himself confesses that when Mars was nearest the Earth, he did not appear much larger than Jupiter which he estimates at only fifty seconds. He errs less, scarcely at all, with regard to Saturn and Jupiter.

But Kepler writes that the proportion of the diameters is without doubt disproved by observation. I reply that he created a shadow which prevented him from seeing clearly. It is true that observation is opposed to it, if his parallax of the Sun, which is of one minute, is to be taken; but I see no necessity for adopting such a parallax, nor do I acknowledge the propriety of his original speculations, much less of his other arguments. Such reasoning is absurd, and like begging the question; the true proportion of the orbits and globes should be sought from observation. In this way the apparent semi-diameter of the Earth, or parallax of the Sun, may be concluded; and if this is borne out by observation the thing is finished.

I say therefore that the diameter of any primary planet, distant from the Sun 15,000 of its own semi-diameters, must appear in the Sun near 0′ 28″ in mean distance. This seems to be consistent with nature; and I will shew in the case of each of the planets that it is not contrary to observation.

1. I will begin with Venus whose diameter I have observed most accurately; and, in her conjunction with the Sun, found to be 1′ 16″, she being, at the time, distant from the Earth 26,409 parts. In her mean distance therefore of 72,333 from the Sun, it appears to be nearly 0′ 28″.

2. The observation which Gassendi made on the 28th of October 1631, proves almost the same thing with respect to Mercury: he found that his diameter in the Sun scarcely equalled twenty seconds. The Rudolphian calculation makes the distance of Mercury from the Earth 67,525; therefore, in his mean distance from the Sun which that calculation states to be 38,806, Mercury will be nearly equal to 0′ 34″, which approaches closely to 0′ 28″, a quantity that is given precisely if four seconds be taken from the

observation, as indeed his words seem to intimate. Thus these two planets preserve their proportion in a remarkable manner, nor do I believe that the rest would differ if they could be observed as carefully; but since we have not the like advantage with regard to them, we must pass on to other methods.

3. Remus and Kepler suppose that Saturn never exceeded thirty seconds, a conjecture which I conceive to be very near the truth, as this planet does not differ perceptibly in respect of distance or diameter. At ten o'clock on the evening of the 6th of September 1639, Saturn appeared as if joined in longitude to a little star placed by Tycho's catalogue in 20° ♑, and he is further said to have appeared at the back of a star of the fifth magnitude, and rather towards the west. The distance compared with the diameter of the moon, was thought to be seven or eight minutes; and upon comparing it afterwards with the diameter of Saturn, I was unable, owing to the great variation, to form a precise estimate; it was however greater than 8 to 1, and less than 16 to 1; Saturn therefore rather exceeded half-a-

minute, but did not equal a whole minute. All this was ascertained by means of a telescope.

4. Kepler supposes (*Astr. Cop.* page 485) that Jupiter covers about fifty seconds by twilight. My proportion gives thirty-seven; the difference is not very great, and may be explained by Jupiter's brightness which increases his appearance. I have often compared Jupiter with Venus, which may be done with certainty, as they shine so equally. On the morning of the 25th February 1640, I thought him rather less; on the 2nd March, I thought him equal or perhaps rather larger; on the 6th, I thought him evidently larger. The diameter of Venus, at that time, was 0′ 24″, according to my estimate; and that of Jupiter about the same quantity. I do not suppose that this calculation is so accurate that a fault of a few seconds may not have arisen in it, either from the variable altitude of the planets, or from the degree of clearness of the diurnal light; but the conjecture is sufficiently satisfactory to my own mind, since it is clear that Jupiter does not differ perceptibly from the proportion of the other planets.

5. The planet Mars loses by comparison with the rest; and certainly does not exceed the assigned proportion. I suppose this is owing to his light being so remarkably obscure, for none of the planets sheds a feebler glow, or diffuses fewer rays. In the beginning of the month of March 1640, Mars appeared much less than Jupiter, though they were in reality equal. He emits however a stronger ray by twilight when he is nearest to the Earth, and sometimes appears so immensely large that he is mistaken by the inexperienced for a new star; on this latter occasion he seems nearly equal to two minutes, a quantity which perhaps he reaches; there is however some doubt upon this point, inasmuch as no other planet, Jupiter and Venus not excepted, actually attains this dimension, though apparently they do not fall far short of it. But there is no need of hesitation when others extend the diameter to six or seven minutes; the proportion here given is at all events probable, and would doubtless agree very well with our observations, if we could make them with sufficient accuracy. It is, without controversy, much more

correct than the opinions put forward by others, which are sometimes many minutes in excess of the truth, as may be seen by referring to the instances of Venus and Mars.

6. Since therefore it is certain that the diameters of the five primary planets, in mean distance, appear from the Sun 0′ 28″, and that none of them deviate from this rule, tell me, ye followers of Copernicus, for I esteem not the opinions of others, tell me what prevents our fixing the diameter of the Earth at the same measurement, the parallax of the Sun being nearly 0′ 14″ at a distance, in round numbers, of 15000 of the Earth's semi-diameters? Certainly, if the Earth agree with the rest as to motion, if the proportion of its orbit to that of the rest be so exact, it is ridiculous to suppose that it should differ so remarkably in the proportion of its diameter. For it is incredible that of the six primary planets the diameter of one should be as much as 2′, or as others make it 6, whilst all the rest should not exceed 0′ 28″. I have not within reach the opinions of other astronomers; but every one must believe what he sees for himself, and to me such a parallax seems absurd.

But it may be replied that this is merely a probable conjecture, and has not the force of demonstration; and further that so immense a distance is unbelievable, inasmuch as it exceeds, by ten times or more, the opinions hitherto received which so many excellent astronomers have geometrically demonstrated from their observations on eclipses. But I answer:

1. I do not put forth this conjecture as an absolute demonstration, but rather as being highly probable, and having as much weight as many others which are carefully received in astronomy. Who, for instance, will prove to me that all the stars are spherical bodies? This has long been known to be true as to the Earth and moon, and has been very recently ascertained as to the Sun and Venus, and the fact that such is the case with them obliges us to suppose, although it cannot be demonstrated by experiments, that it is so with Jupiter and Saturn, &c.; at all events they are not planes as they appear to be to us. Kepler rightly concluded that the figure of the orbits of all the planets is elliptical; and though this cannot be verified with respect to

Venus and the Earth, on account of their small eccentricity, it is sufficient that observations do not disprove in their case that form which is required in the case of others, and it is enough that no good reason can be alleged why we should not assign to the Earth the same proportion which all other planets possess.

2. It has lately been shewn, from the diameter of Venus, how little importance is to be attached to the common opinion of astronomers respecting the Sun's parallax; for though the planet was so long open to observation, and her diameter could have been measured by so many different methods, it is fixed, by common consent, at least ten times as great as it ought to be. What fear then of innovation can arise from my stating that the same thing has happened in respect of the diameter of the Earth, the appearance of which in the Sun no one ever saw, and the investigation of which is most difficult, and has not hitherto been properly undertaken?

3. Moreover if any one has clearly demonstrated from observation a greater parallax, and does not find mine to be in all respects confirmed,

I am willing to reject it as a false speculation. I know how loudly some speak of the distance of the Sun as demonstrated from the centre of the Earth; but they are triflers, seeking for vain glory, and trying to impose fallacies upon the credulous, instead of bringing forward actual proof.

I had intended to offer a more extended treatise on the Sun's parallax; but as the subject appears foreign to our present purpose, and cannot be dismissed with a few incomplete arguments, I prefer discussing it in a separate treatise, "*De syderum dimensione*" which I have in hand. In this work, I examine the opinions and views of others; I fully explain the diagram of Hipparchus by which the Sun's parallax is usually demonstrated; and I subjoin sundry new speculations; I also shew that the hypotheses of no astronomer, Ptolemy not excepted, nor even Lansberg who boasts so loudly of his knowledge of this subject, answer to that diagram, but that Kepler alone properly understood it; I shew in fact that the hypotheses of all astronomers make the Sun's parallax either absolutely nothing, or so small that it is quite imperceptible, whereas they themselves, not understanding what they are about, come to an entirely opposite conclusion, a paradox of which Lansberg affords an apt illustration. Lastly, I shew the insufficiency and uselessness of the common mode of demonstration from eclipses; I give many other certain and easy methods of proving the distance and magnitude of the Sun, and I do the same with regard to the moon and the rest of the planets, adducing several new observations.

London:—Wertheim, Macintosh, and Hunt, 24, Paternoster Row, and 23, Holles Street, Cavendish Square.

Printed in the United States
By Bookmasters